journey into power

TEN AREAS OF STUDY

chelsea selvadurai

ISBN 978-0-359-88222-9

the move on

Mission

Exist as an uplifting resource for those affected by abuse, mental illness, loneliness, and the weariness caused by modern life. To promote intentional living through personal essays of the physical; exploration and exercise, literal; by way of relocation, and spiritual; moving away from things that bring harm.

My hope is that this research booklet, created as part of my 200-hour yoga teacher certification, continues the work of my personal mission, that information herein guides those who wish to use their lights to illuminate the world, that they may always keep moving onwards, upwards, intentionally, and brilliantly.

Namaste.

Chelsea Selvadurai
themoveon.org

journey into power
THE SEQUENCE

INTEGRATION

AWAKENING

VITALITY

EQUANIMITY

GROUNDING

IGNITING

STABILITY

OPENING

RELEASE

REJUVENATION

DEEP REST

integration

CHILD'S POSE

DOWNWARD FACING DOG

STEP FORWARD/HALFWAY LIFT

RAGDOLL

MOUNTAIN

SAMASTHITI

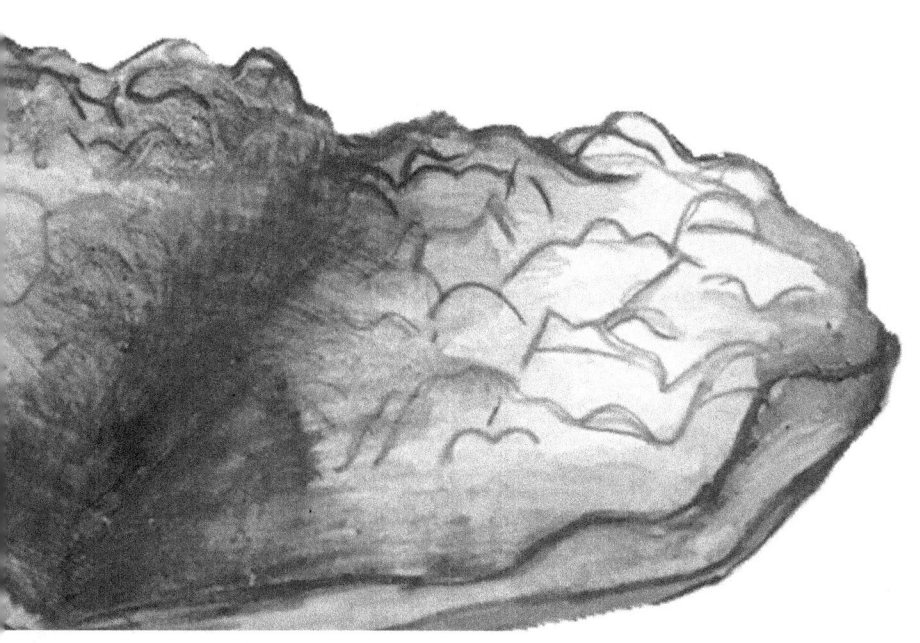

child's pose
(BALASANA)

GENERAL FORM
Beginning on the floor, knees are widened to the outer edge of the mat while big toes come to touch.
Hips shift towards the buttocks and arms extend overhead, palms facing down. Forehead comes to the mat.

COMMON MISALIGNMENTS
Inability to relax onto the mat due to stiff knees, hips, back, or tight quads.
Diaphragm compressed or throat closed.

CONTRAINDICATIONS & MODIFICATIONS
Those with knee problems may find this pose uncomfortable. Knees can be drawn together to alleviate weight.
Arms can be rested alongside the torso with palms facing up, releasing shoulders toward the floor.

BIOMECHANICS
Settles the nervous system and squeezes internal organs.

VARIATIONS & ADVANCEMENTS
Arms can be rested alongside the torso with palms facing up, releasing shoulders toward the floor.

BENEFITS
While deeply relaxing, child's pose stretches the lower back and inner thighs, promoting flexibility and circulation in
the muscles, joints, and discs of the back.

ENERGETICS
In the first pose of the sequence, a seasoned yogi may observe the physical connection to their mat as an invitation
to disconnect from the outside world. This calming freedom and anticipation promotes intentional breathing and
thoughts, creating the foundation for the commitment to follow.

CUES
Reach fingers towards the front of the mat and drop your hips. Begin to bring intention to your breath.

ASSISTING
Light pressure on the lower back, fists in the arches of feet.

SANSKRIT TRANSLATION & STORIES
Balasana (buh-LAHS-ahna) comes from the Sanskrit *bala* "child" and *asana* "posture". During Krishna's childhood,
he displayed a playful forgetfulness of his own divinity (*lila*), which allowed intimacy between the contingent world
and the infinite. Balasana invites the practitioner to surrender of the ego and give selfless love – two actions that
children do best.

downward facing dog
(ADHO MUKHA SVANASANA)

GENERAL FORM
The general form of downward facing dog is a triangle formed with the body with hands and feet on the mat.

COMMON MISALIGNMENTS
Hands and feet are not centered or firmly rooted to the floor, stressing the wrists or ankles, too little space between hands and feet, upper back rounded, head lifting or neck craning, elbows locked out.

CONTRAINDICATIONS & MODIFICATIONS
Those with wrist, shoulder or carpal tunnel injuries may find this pose uncomfortable.
Puppy, Dolphin, and Four Post poses may alleviate discomfort.

BIOMECHANICS
First inversion of the class will stimulate the cardiovascular system. The psoas, pectineus, sartorius, and rectus femoris combine to flex the hips and trunk. The abdominals contract to draw the internal organs inward.

VARIATIONS & ADVANCEMENTS
Three-Legged Dog, Puppy, and Dolphin poses.

BENEFITS
Builds strength in the hands, wrists, lower back, hamstrings, calves, and Achilles tendon while elongating the back and shoulder area. It can be an active pose for stretching and strengthening various regions of the body, or a pose in which to actively rest during practice.

ENERGETICS
Second to Child's Pose, Downward Facing Dog invites a "coming home" feeling while awakening the entire body. It is the place to check in, to return to breath, adjust to proper alignment, and heighten the awareness of increased flexibility and strength throughout the practice.

CUES
With hands shoulder width apart, spread your fingers wide and align your wrist crease to the front of the mat. Emphasize pressing down the index and thumb pads. Upper arms externally rotate. Engage the lower belly by drawing the navel towards the spine. Send the sit-bones and tailbone up and back. Inner thighs rotate inwards as you firm the outer thighs.

ASSISTING
Facing the student, light pressure on the low back.
Behind the student, light pressure "clawing" the Achilles, grounding heels towards the earth.

downward facing dog continued...

Sanskrit Translation & Stories

The name comes from the Sanskrit *adhas* "down", *mukha* "face", *svana* "dog", and *asana* "posture". One myth of Adho Mukha Svanasana (AH-doh MOO-kah shvah-nahs-anna) stems from a Vedic hymn where Sarama, a female dog of the gods, was sent to find the sacred cows of a sage stolen by a rival tribe. Loyal in nature, Sarama successfully retrieved the livestock after withstanding threats of violence and bribery, having said to have located the herd through "the path of the Truth". In one hymn, this act is described as "the Dawn who recovers the rays of the Sun that have been carried away by night".

halfway lift
(ARDHA UTTANASANA)

GENERAL FORM
Feet are parallel and upper body is elongated perpendicular to the legs,
hands press into shins or floor to create a long spine.

COMMON MISALIGNMENTS
Neck or shoulders "crunched" up, abdominal muscles not engaged, chin lowered, back arched.

CONTRAINDICATIONS & MODIFICATIONS
Tight back or hamstrings. Modify by placing a block under hands to bring the earth closer, or bend the knees.

BIOMECHANICS
Spine is lengthened to make room for spinal fluid to flush.

VARIATIONS & ADVANCEMENTS
Knees bent.

BENEFITS
Lengthens and strengthens the spine, strengthens and stretches the hamstrings, tones the obliques, and prepares for the next pose.

ENERGETICS
Reset and attention to breath and alignment. Preparedness.

CUES
Inhale as your lift your upper torso to create a flat back. Press your seat back and pull your chest forward as you elongate the spine. Allow your fingertips to guide your lift and lengthen the back of your neck as you gaze forward. Pull belly up and in.

ASSISTING
Facing the side body, one hand applies light pressure to the sacrum, while the other hand traces the spine.

SANSKRIT TRANSLATION & STORIES
The root of Ardha Uttanasana (ARD-ha OOH-ton-AH-san-ah) is *ardha*, "half", descending from the androgynous Hindu deity, Ardhanarishvara. Ardhanarishvara's ability to embody both masculine and female energies (*Pursha & Prakriti*) demonstrates how Shakti, the female principle of God, is inseparable from (and equal to) Shiva, the male principle of God.

ragdoll
(UTTANASANA)

GENERAL FORM
Upper body is folded over lower body while hands grasp elbows.

COMMON MISALIGNMENTS
Knees locked out, feet pointing inwards or outwards, weight sinking into heels, neck not relaxed.

CONTRAINDICATIONS & MODIFICATIONS
Those with tight hamstrings, back or next injury may find this pose uncomfortable.
Modify by widening your stance, bending knees, or placing hands on blocks.

BIOMECHANICS
The psoas, pectineus, and rectus femoris flex the hips and tilt the pelvis slightly forward. Quadriceps contract to straighten knees, creating reciprocal inhibition, relaxing the hamstrings. Shoulders draw away from the neck.

VARIATIONS & ADVANCEMENTS
Bend into the knees and take a twist, bind arms behind legs, or sway side to side.

BENEFITS
Ragdoll strengthens the spine and thighs, improves posture and relieves stress.

ENERGETICS
Simplicity, preparedness, grounded, and ready.

CUES
Feet are hips width (or wider) with toes facing 12 o'clock. Bend your knees as much as needed as your torso folds forward, elongating your spine. Relax your neck and let the upper body hang.

ASSISTING
Facing the student from the side, medium pressure can be applied on the low back, and option for the opposite hand to apply pressure to the spine down to the base of the neck.

SANSKRIT TRANSLATION
The direct translation of the uttanasana (OOH-ton-AH-san-ah is *ut* "intense", *tan* "to stretch, extend, or lengthen out", and *asana*, "posture".

mountain
(TADASANA)

GENERAL FORM
Standing upright with feet hip width distance and at 12 o'clock. Arms reach up, palms facing each other.

COMMON MISALIGNMENTS
Feet not at True North, pelvis tipped forward, shoulders near ears, excessive arch in back.

CONTRAINDICATIONS & MODIFICATIONS
Shoulder injury, low blood pressure, headache.
Modify by bringing arms to prayer or cactus.

BIOMECHANICS
Muscles on the top and bottom of the feet balance each other, grounding the pose. Calf muscles work quietly to balance the ankles. Alignment is neutral.

VARIATIONS & ADVANCEMENTS
Samasthiti (Prayer Pose).

BENEFITS
Improves posture, strengthens thighs, and stimulates nervous system.

ENERGETICS
Tadasana invites a connection to the presence.
Ray Long in The Key Poses of Yoga quips: "It is as though we have climbed a plateau to gauge the transformative effects of our practice and collect our muscle-thoughts before continuing our ascent."

CUES
Root into the ground through your feet. Inhale, lift your chin and sweep hands overhead. Bring energy into your fingertips as palms face each other. Tuck your tailbone and knit ribs together.

ASSISTING
Hands guide the forearms as they rise. Light pressure on each side of the neck to pull shoulders away from ears.

SANSKRIT TRANSLATION
Tadasana (tuh-DAHs-ahna) is *tada* "mountain" and *asana* "posture".

samasthiti
(POSE CALLED IN SANSKRIT)

GENERAL FORM
Standing upright with feet hip width and at 12 o'clock, palms are together at heart center and gaze is forward.

COMMON MISALIGNMENTS
Feet not at True North, pelvis tipped forward, shoulders near ears, excessive arch in back.

CONTRAINDICATIONS & MODIFICATIONS
Wrist injury, arthritis. Arms at side.

BIOMECHANICS
Alignment is neutral.

VARIATIONS & ADVANCEMENTS
Arms at side.

BENEFITS
Improves posture, strengthens thighs, and stimulates nervous system.

ENERGETICS
Attention is centered on the present and focused on the intentional.

CUES
Use your feet to root into the ground. Elongate your neck and allow the crown of your head to rise to the sky, bringing shoulders away from ears. Breathe, and set an intention for your practice.

ASSISTING
Light pressure on either side of the neck to pull shoulders away from ears.

SANSKRIT TRANSLATION & STORIES
The name comes from the Sanskrit *sama* "same" and *sthiti* "to establish". Similar to Pranamasana or Anjali Mudra, a pose of intention and respect. While the ancient word *sthiti* has been subject to multiple interpretations, the Vedic origin of sthiti refers to the one reality which exists in wholeness, possesses multiple forms and qualities, and upholds the world and all within - whatever desires to move or remain stationary, whatever is breathing or not breathing, whatever is seeing or unseeing. The word itself refers to maintenance, preservation, and continued existence in any place with a sense of position and permanence.

awakening
(SUN SALUTATION A)

MOUNTAIN
FORWARD FOLD
HALFWAY LIFT
HIGH PLANK
LOW PLANK
UPWARD FACING DOG
DOWNWARD FACING DOG

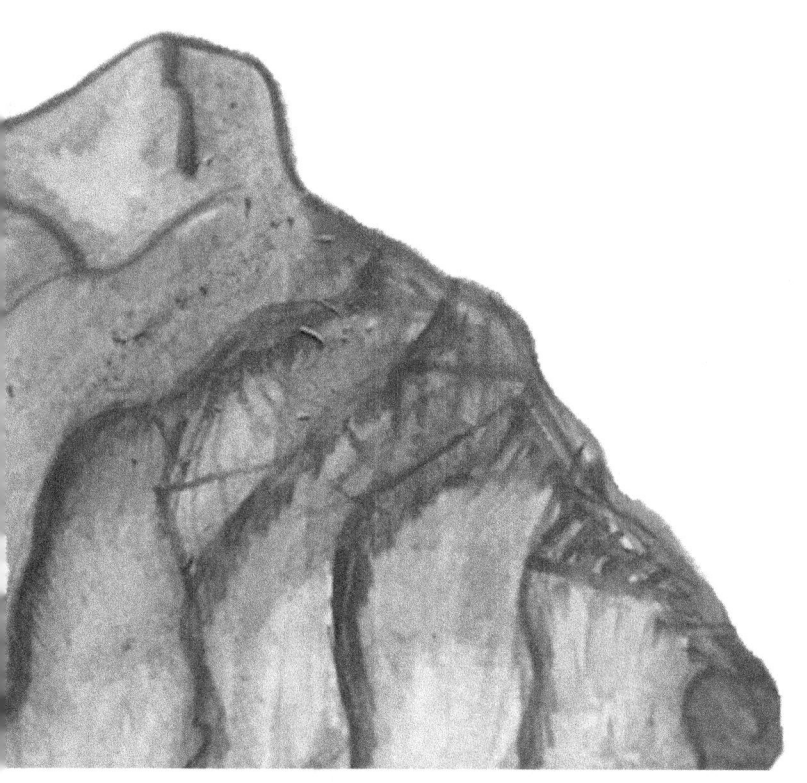

awakening
(SUN SALUTATION B)

CHAIR

FORWARD FOLD

HALFWAY LIFT

HIGH PLANK

LOW PLANK

UPWARD FACING DOG

DOWNWARD FACING DOG

WARRIOR ONE, RIGHT SIDE

HIGH PLANK

LOW PLANK

UPWARD FACING DOG

DOWNWARD FACING DOG

REPEAT SEQUENCE ON OPPOSITE SIDE

forward fold
(UTTANASANA)

GENERAL FORM
Upper body is folded over lower body, attempting to touch toes.

COMMON MISALIGNMENTS
Knees locked out, feet not parallel, weight sinking into heels, neck not relaxed.

CONTRAINDICATIONS & MODIFICATIONS
Those with tight hamstrings, back or next injury may find this pose uncomfortable.
Modify by widening your stance, bending knees, or placing hands on blocks.

BIOMECHANICS
The psoas, pectineus, and rectus femoris flex the hips and tilt the pelvis slightly forward. Quadriceps contract to straighten knees, creating reciprocal inhibition, relaxing the hamstrings. Shoulders draw away from the neck.

VARIATIONS & ADVANCEMENTS
Widening stance, blocks under hands.

BENEFITS
Stretch in lower back and hamstrings.

ENERGETICS
Grounding, while bringing focus and diagnostics to the lower back.

CUES
Exhale and fold forward from the waist, hinging at your hips. Back stays flat and abdominal tucks towards spine. Bend your knees as much as needed as your torso folds forward, elongating your spine. Arms reach towards the floor. Allow your head to hand down and your neck to be free.

ASSISTING
Facing the student from the side, medium pressure can be applied on the low back, and option for the opposite hand to apply pressure to the spine down to the base of the neck.

SANSKRIT TRANSLATION
The direct translation of Uttanasana (OOH-ton-AH-san-ah is *ut* "intense", *tan* "to stretch, extend, or lengthen out", and *asana*, "posture".

high plank
(DANDASANA)

GENERAL FORM
Torso is parallel to the mat with arms straight, hands flat on the mat and facing forward. Legs are outstretched with big toes, ankles, and knees parallel to each other.

COMMON MISALIGNMENTS
Locked elbows, hands in front of shoulders, or sagging back, abdominals and/or shoulder blades.

CONTRAINDICATIONS & MODIFICATIONS
Shoulder, wrist, or lower back injury, and carpal tunnel syndrome. Knees can be lowered to the floor.

BIOMECHANICS
The rectus abdominis tightens to oppose the back muscles and holds the trunk solid. Calf muscles flex the ankles and push off from the feet, while the force of triceps and deltoids counters this action, helping stabilize the pose.

VARIATIONS & ADVANCEMENTS
Forearm plank, plank pushups.

BENEFITS
Strengthens core, arms, and wrists.

ENERGETICS
A powerful strength pose, adding additional intention to alignment prior to Chaturanga Dandasana.

CUES
Squeeze quadriceps and buttock muscles and draw in your front ribs.
Stack your shoulders over your elbows and wrists.

ASSISTING
Straddling the student from behind, medium pressure on either hip aids in centering through the vinyasa.

SANSKRIT TRANSLATION & STORIES
If you think holding high plank (don-DAHS-ahna) feels like a punishment... maybe here's why.
Danda translates to "stick" (or staff), a symbol of authority in Hindu culture. The term can be visibly evident as a scepter (or strong, straight arms and legs in the pose), but also considered emblematic of the kings' social obligation to maintain social order by inflicting punishment on evil doers.

low plank
(CHATURANGA DANDASANA)

GENERAL FORM
Plank with arms lowered to a right angle.

COMMON MISALIGNMENTS
Elbows out (not hugging the body), arched back, shoulders rounded, misalignment in high plank.

CONTRAINDICATIONS & MODIFICATIONS
Shoulder, wrist, or lower back injury, carpal tunnel syndrome, or pregnancy. Knees can be lowered to the floor.

BIOMECHANICS
The rectus abdominis tightens to oppose the back muscles and holds the trunk solid. Calf muscles flex the ankles and push off from the feet, while the force of triceps and deltoids counters this action, helping stabilize the pose.

VARIATIONS & ADVANCEMENTS
Roll over balls of feet to the tops of the feet and shift torso forward, bringing hands back by waist to increase challenge.

BENEFITS
All over strengthening pose, specifically arms, wrists, and abdominals.

ENERGETICS
A powerful strength pose, adding additional intention to alignment.

CUES
From high plank, exhale and bend elbows 90 degrees, keeping them tucked into your sides, and move forward as your torso lowers. Keep legs and core activated.

ASSISTING
Straddling the student from behind, medium pressure on either hip aids in centering through the vinyasa.

SANSKRIT TRANSLATION & STORIES
While Chaturanga Dandasana (chat-uhrr-UNG-uh don-DAS-ahna) simply means "four-limbed" and "staff", the use of the word chaturanga and its four divisions attributes to battle strategies in Indian epic poetry. In *Mahabharata* (400 BC), the four parts are elephantry, chariotry, cavalry, and infantry. Chaturanga also lends its name to the ancestor of chess (developed in 6th century AD), where an 8x8 checkerboard mimics ancient battle formations featuring the four components.

upward facing dog
(URDHVA MUKHA SVANASANA)

GENERAL FORM
Lying facing the mat, the upper body pushes up, pressing hands and feet against the mat while torso and hips rise into a backbend.

COMMON MISALIGNMENTS
Collapsing into low back, hands in front of shoulders, shoulders by ears, craned neck, locked out arms, quadriceps not engaged and thighs not lifted from mat.

CONTRAINDICATIONS & MODIFICATIONS
Back injury, carpal tunnel syndrome, pregnancy.
Modify by bending elbows and tucking into sides, while thighs come to the mat.

BIOMECHANICS
Muscles on the back of the shoulder blades turn the shoulders outward to open the chest.
Glutes and abdominals stabilize the pelvis and protect the lower back.

VARIATIONS & ADVANCEMENTS
Knees lift off floor. Cobra pose.

BENEFITS
Strengthens upper extremities, opens the chest, and tones the extensors of the back.

ENERGETICS
Chest and heart are open, inviting a sense of personal power and vulnerability.

CUES
From low push up, with wrists directly under shoulders, press down through your hands to lift your torso, scooping your belly up as you straighten your arms. Push down through the tops of your feet and engage your legs and glutes to lift thighs off of the mat. Gaze is forward.

ASSISTING
Parallel to assist in high/low plank, hands are placed firmly on the outside of the shoulder/upper arms.

SANSKRIT TRANSLATION & STORIES
The name of the pose is from the Sanskrit *urdhva* "upwards", *mukha* "face", *shvana* "dog", and *asana*, "posture".

chair
(UTKATASANA)

GENERAL FORM
From standing, knees bend as if sitting back into a chair, arms extend overhead.

COMMON MISALIGNMENTS
Weight shifted onto toes or shifting, knees knocking into one another, shoulders up near ears.

CONTRAINDICATIONS & MODIFICATIONS
Knee, lower back, or shoulder injury. Modify by reducing bend in the knees for lower body relief, and/or bringing hands to heart center or cactus for upper body relief.

BIOMECHANICS
The hip flexors hold the femurs in a slightly flexed position, the psoas provides a counterbalance to the lower back muscles, aiding to protect the lumbar spine, and the middle trapezius and rhomboids combine to draw the shoulder blades towards the midline of the back and open the chest.

VARIATIONS & ADVANCEMENTS
Raise onto the balls of your feet as you are lowering.

BENEFITS
Chair Pose strengthens several core muscle groups, including those that flex the pelvis, the quadriceps, and lower back muscles.

ENERGETICS
The pose suggests potential energy to be unleashed, like a thunderbolt!

CUES
Feet are at 12 o'clock. Lift toes and shift weight into heels. Draw pelvis downward as you bend your knees, maintaining a neutral spine and strength in your legs. Broaden your upper back and drop shoulders down and away from your ears.

ASSISTING
Medium pressure applied to each hip, or tops of shoulders.

SANSKRIT TRANSLATION & STORIES
Utkatasana's (OOHT-kuh-TAS-ahna) root word of *utkat* has many interpretations from strong to wild, frightening, above all the usual, intense, gigantic, furious, and heavy. Chair pose is also sometimes referred to as "thunderbolt" due to the positioning of the body.

warrior one
(VIRABHADRASANA ONE)

GENERAL FORM

Lunge with front knee bent at 90 degrees while back leg is straight. Shoulders, hips, and front foot points at 12 o'clock and back foot at a 45-degree angle. Arms are raised overhead with palms facing each other.

COMMON MISALIGNMENTS

Not grounded in outer edge of back foot, front knee not stacked over ankle or collapsing inward. Hips or shoulders not parallel to top of mat.

CONTRAINDICATIONS & MODIFICATIONS

Groin strain, lower back or shoulder injury. Modify by shortening the stance and depth of bend in the front knee. Hands can be placed at heart center, cactus, or hips.

BIOMECHANICS

The buttock muscle of the back leg extends and turns the hip outward. Balance is assisted by the sartorius bending the hip and turning the thigh outward. The core is engaged to contract slightly, protecting the lower back.

VARIATIONS & ADVANCEMENTS

Hands can be placed at heart center, cactus, or hips, or clasped behind torso.

BENEFITS

Increases leg strength and hip flexibility. Opens chest and works the back muscles.

ENERGETICS

Powerful, disciplined.

CUES

Lunge forward and aim for a 90-degree angle. Press into the outer edge of your back foot. Square your hips. Lift your heart up and set your drishti forward.

ASSISTING

Medium pressure applied to each hip, or tops of shoulders.

SANSKRIT TRANSLATION & STORIES

Virabhadrasana (VEER-uhb-hahd-DRA-sahna) comes from the warrior Virabhadra (also known as Veerabathira, Veerabathiran, Veeraputhiran) who is an extremely fierce and fearsome form of the Hindu god Shiva. Virabhadrasana I is Shiva's arrival, with swords in both hands thrusting his way up through the earth. (Read more about the myth of Virabhadrasana in Warrior Two).

vitality

REVOLVING CRESCENT LUNGE
WARRIOR TWO
EXTENDED SIDE ANGLE
SIDE PLANK

HIGH PLANK : LOW PLANK : UPWARD FACING DOG : DOWNWARD FACING DOG

REPEAT OTHER SIDE

STEP FORWARD : HALFWAY LIFT : FORWARD FOLD

CHAIR
PRAYER TWIST, RIGHT
BIG TOE POSE
CHAIR
PRAYER TWIST, LEFT
GORILLA
CROW

VINYASA TO DOWNWARD FACING DOG
STEP FORWARD : HALFWAY LIFT :
FORWARD FOLD : MOUNTAIN

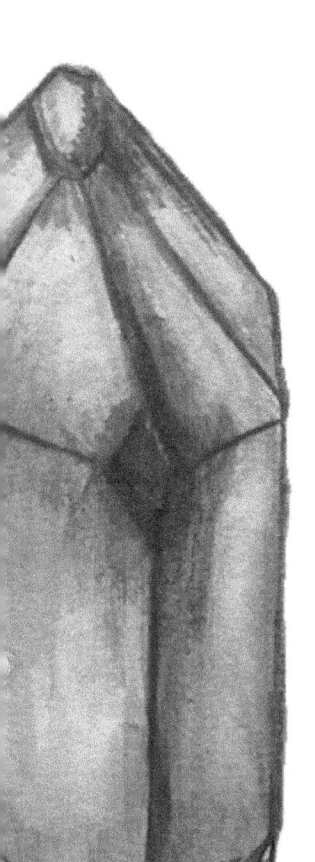

crescent lunge
(ANJANEYASANA)

GENERAL FORM
From true north, take a large step forward. Front knee lowers into a 90-degree lunge while the back heel lifts, with both feet facing forward. Arms are stretched overhead, palms facing each other.

COMMON MISALIGNMENTS
Overextended backbend, shoulders elevated to the ears, back foot not pushed into the ground.

CONTRAINDICATIONS & MODIFICATIONS
Avoid if you have high blood pressure or knee injury. Hands can be rested on thighs for shoulder injuries. Back knee may lower to the mat with toes untucked.

BIOMECHANICS
Quadriceps straighten the knee in the rear leg, the front leg bends at the hip while shortening the psoas and pectineus. The pose is balanced by the Sartoris bending the hip, turning the thigh outward.

VARIATIONS & ADVANCEMENTS
Crescent Moon (back knee down), incorporating a bind by interlacing hands behind your back and opening the chest, closing the eyes and focusing on balance.

BENEFITS
Opens the hips and chest, lengthens the spine and strengthens the lower body. Promotes flexibility into the front and back of the torso. Improves circulatory system, balance, and core awareness.

ENERGETICS
Calms the mind. Powerful and strong.

CUES
Palms face each other with energetic fingertips reaching towards the ceiling. Knee bent to 90 degrees, careful not to extend past the ankle. Strong in your back leg. Both feet facing forward, core engaged. Tuck tailbone.

ASSISTING
Palms are placed on hips. If the yogi is shaky, you can gently squeeze the back leg with your legs to stabilize.

SANSKRIT TRANSLATION & STORIES
Anjaneyasana's (AHN-juh-nay-AAH-suh-nuh) root word of *anjaneya* translates to "son of Hanuman's mother, Anjana". When Anjaneya was a boy, Anjana told him that all red fruit was his to eat. Looking upon the sun, he thought by eating it he would never be hungry again. As he reached for the sun, a monster seemed to be swallowing the fruit he was after, which unbeknownst to him was a solar eclipse (crescent!). Thus, this pose resembles a young child reaching towards the sky.

revolving crescent lunge
(PARIVRTTA ALANASANA)

GENERAL FORM

From crescent lunge (90-degree lunge, arms overhead, and strong back leg) hands come to heart center and shoulders lower towards the lunged leg, hooking the elbow outside of the front knee.

COMMON MISALIGNMENTS

Arms contact bent knee, back foot not engaged, chest collapsed, gaze down.

CONTRAINDICATIONS & MODIFICATIONS

Those with knee or spinal injuries should avoid. Knee can be lowered to the mat.

BIOMECHANICS

Creates a stretch of the core muscles surrounding the spine. The pose is stabilized by activating the front leg psoas and back leg glutes. Internal organs, glands, and circulatory system is activated.

VARIATIONS & ADVANCEMENTS

Arms open, side crow, bind with arms behind the back with hands interlacing.

BENEFITS

Improves internal organ health, glands, and circulatory system by flushing away the blood and, upon release, brings new oxygen rich blood to the organs and muscles, washing away toxins and aiding in digestion.

ENERGETICS

Deep concentration, Strength. Confidence.

CUES

Press hands together to continue deepening the stretch. Press high onto the ball mound of your foot. Pull belly in and shoulder blades back. Reach the crown of your head forward to lengthen the spine.

ASSISTING

Gently place palm of hands on hips.

SANSKRIT TRANSLATION & STORIES

Para comes from the Sanskrit *parsva* "side". *Vrtta* refers to a unit of limited measure, as in "revolves to a certain point" vs. spinning or continuously linear, and *asana* "posture".

warrior two
(VIRABHADRASANA TWO)

GENERAL FORM

Front leg is lunged into 90-degrees and rear leg is straight behind with toes parallel to the short edge of the mat. Shoulders are relaxed and aligned with hips, while arms are parallel to the floor reaching away from the midline of the body. Gaze is forward over front hand.

COMMON MISALIGNMENTS

Base is not wide enough or too wide, heels not aligned, shoulders raised into ears, front knee lunging over toes or buckling in/out, arms not in line with legs or limp.

CONTRAINDICATIONS & MODIFICATIONS

Groin strain, ankle, shoulder, or knee injuries.
Stance can be shortened and degree of knee bend can be reduced. Hands can come to heart center.

BIOMECHANICS

The rear leg buttock muscles extend and externally rotate the hip while the quadriceps straighten the back knee. In the core, the rectus abdominis contracts to protect the lower back. In the upper body the deltoids raise the arms and draw them slightly backward to open the chest.

VARIATIONS & ADVANCEMENTS

This pose is challenging enough to stay aligned in the deepest expression!
Arms can be varied with binds or cactus arms.

BENEFITS

Warrior Two is a hip opening pose that sculpts the muscles in the buttocks and legs.

ENERGETICS

A pose of incredible mental focus, one may also feel strong and empowered.

CUES

Aim to have your front thigh parallel to the floor while working to press through the outer edge of your back foot, stretching your mat apart. Shoulders relax and remain aligned over hips. Engage the arms by inviting energy into your fingertips. Pull your belly button into your spine and gaze forward, chin level to the mat.

ASSISTING

Hands may come to the side of the neck to encourage shoulders to relax. Attention to the back foot or hips for proper alignment may be necessary.

SANSKRIT TRANSLATION & STORIES

Virabhadrasana (VEER-uhb-hahd-DRA-sahna) is named for the warrior Virabhadra. The powerful priest Daksha held a great ritual sacrifice (*yagna*) but did not invite his youngest daughter, Sati and her husband Shiva, the supreme ruler of the universe. Sati found out and decided to go alone. When she arrived, Sati entered into an argument with her father. Unable to withstand his insults, she spoke a vow to her father, "Since it was you who gave me this body, I no longer wish to be associated with it." She walked to the fire and threw herself in. When Shiva heard of Sati's death, he was devastated. He yanked out a lock of his hair and beat it into the ground, where up rose a powerful Warrior. Shiva named this warrior Virabhadra and ordered him to go to the yagna and destroy Daksha and all his guests. In Virabhadrasana II, he has Daksha in his sights. In Virabhadrasana III, he decapitates Daksha with his sword.

extended side angle
(UTTHITA PARSVAKONASANA)

GENERAL FORM
Keeping the legs in the same form as Warrior two, the torso drops towards the lunged leg and twists open, the top arm reaching towards the ceiling. The arm closest to the bent leg may come to a block, the floor, or the elbow gently pressing into the inside of the knee.

COMMON MISALIGNMENTS
Using the bent leg too aggressively for support. Coming out of the front lunge or lifting the back heel. Ribs not spiraling up, rounding of the back.

CONTRAINDICATIONS & MODIFICATIONS
Groin, knee, back, or shoulder injury.

BIOMECHANICS
Strengthens and stretches the legs, knees, and ankles, stretches the groins, spine, waist, chest and lungs, and shoulders, stimulates abdominal organs and increases stamina

VARIATIONS & ADVANCEMENTS
Arms may come to a bind, or the arm closest to the bent leg may reach forward to engage the core.

BENEFITS
Stimulates abdominal muscles and aids in rinsing toxins from the organs.
Opens the chest and strengthens the entire body.

ENERGETICS
Feeling of strength, stability, and balance.

CUES
Continue to aim for your front leg to be parallel to the floor. Scoop your tailbone down and spiral your rib cage towards the ceiling. Feel the side stretch of your body from your back leg to the top of your fingers.

ASSISTING
Align foot to foot (back leg) and gently glide the top hand upwards to promote the opening of the chest

SANSKRIT TRANSLATION & STORIES
Utthita Parsvakanasana (OOT-hee-tuh PAR-svuh-kuh-NAH-sun-uh) comes from
the Sanskrit *utthita* "extended", *parsva* "side or flank", *kona* "angle", and *asana* "posture".

side plank
(VASISTHASANA)

GENERAL FORM

From high push up, spin to the outside of one foot while hips stay stacked. The top hand comes to the ceiling, bottom arm straight and strong in one line.

COMMON MISALIGNMENTS

Weight may easily fall entirely to the wrist by not activating the core. Hips may be too high or too low.

CONTRAINDICATIONS & MODIFICATIONS

Shoulder, wrist, or elbow injuries. The lower knee can be lowered to the ground.

BIOMECHANICS

Strengthens the muscles that influence frontal plane biomechanics and alignment. Primary muscles are hip adductors, abductors, and internal and external obliques. Frontal plane stability, or lack thereof, plays a large role in stance phase mechanics and facilitating proper alignment in hips, knees, feet, and ankles.

VARIATIONS & ADVANCEMENTS

"Wild Thing". Top leg can be lifted to the ceiling, can even grab the upper leg/toe with top hand.

BENEFITS

Strengthens abdominals and stabilizes the wrist and elbows. Integrates the upper and lower body.

ENERGETICS

This is a pose of true strength, with room for childlike play in its many variations.

CUES

Keep heels, hips and heart in one line. Create a line of pure energy flowing from your raised hand all the way into the mat. Pull your belly button into your spine as hips lift. Gaze up at your top hand.

ASSISTING

Similar to Utthita Parsvakonasana, the top hand can be glided upwards, granted that the foundation of the side plank is strong.

SANSKRIT TRANSLATION & STORIES

The name comes from the Sanskrit *Vasistha* (a sage), and *asana* "posture". Vashishtha is Sanskrit for "most excellent, best, or richest", and was one of the oldest and most revered Vedic rishis and one of the Saptarishis (seven great Rishis) of India. His ideas have been influential and he was called as the first sage of the Vedanta school of Hindu philosophy.

prayer twist
(PARIVRTTA UTKATASANA)

GENERAL FORM
Chair pose, with hands at heart center, twisting the torso to one side.

COMMON MISALIGNMENTS
Knees or hips coming out of alignment.

CONTRAINDICATIONS & MODIFICATIONS
Lower back injuries. Lower hand to a block or gaze to the floor.

BIOMECHANICS
The quadriceps are active, holding the knees in partial flexion. Obliques are engaged through the rotation of the upper body.

VARIATIONS & ADVANCEMENTS
Arms can be open to the side, bound, or many yogis take side crow.

BENEFITS
Creates flexibility in the middle to lower back, and offers a rinse of organs and muscles of midsection, detoxifying organs and glands.

ENERGETICS
A pose of strength, building upon an already challenging pose. Continued energy through lengthening.

CUES
Place your elbow on the outside of your thigh and work to keep hands at heart center. Inhale as spine lengthens forward, exhale as you twist deeper. Continue to work towards equal balance in both feet and keeping knees and hips aligned towards the front of your mat.

ASSISTING
Attention to the squaring of hips, or bring attention to the shoulders creeping up towards ears by applying light to medium pressure.

SANSKRIT TRANSLATION & STORIES
Para comes from the Sanskrit *parsva* "side". *Vrtta* refers to a unit of limited measure, as in "revolves to a certain point" (vs. spinning or continuously linear), and *asana* "posture".
Utkatasana's root word of *utkat* has many interpretations from strong to wild, frightening, above all the usual, intense, gigantic, furious, and heavy.

big toe pose
(PADANGUSTHASANA)

GENERAL FORM
A forward fold, holding onto the big toe with "peace fingers" while working to pull the upper body forward.

COMMON MISALIGNMENTS
Feet not aligned at "true north", knees locked out, straining the neck, forcing a forward bend.

CONTRAINDICATIONS & MODIFICATIONS
Lower back injuries. Feet may be placed wider, with bent knees or blocks as needed.

BIOMECHANICS
Releases hamstring and lower back.

VARIATIONS & ADVANCEMENTS
Use of yoga strap to continually bring the upper body towards the lower body.

BENEFITS
Strengthens the thighs and hamstrings while providing an active recovery from challenging poses.

ENERGETICS
An active relief between chair poses. With continued effort, an empowering pose promoting flexibility.

CUES
Toes point at twelve o'clock. Inhale to lengthen the spine, exhale and sink deeper into the pose. Neck is relaxed and reaches towards the mat. Elbows reach out wide.

ASSISTING
Facing the student from the side, medium pressure can be applied on the low back, and option for the opposite hand to apply pressure to the spine down to the base of the neck.

SANSKRIT TRANSLATION & STORIES
The name is from the Sanskrit *padangustha* "big toe" and *asana*, "posture".
The asana is not described in medieval hatha yoga texts, but appears in the 20th century.

gorilla
(PADAHASTASANA)

GENERAL FORM
Forward fold with hands under feet, toes coming to the wrist crease.

COMMON MISALIGNMENTS
Feet not aligned at "True North", knees locked out, straining the neck, forcing a forward bend.

CONTRAINDICATIONS & MODIFICATIONS
Lower back injuries. Feet may be placed wider, with bent knees or blocks as needed.

BIOMECHANICS
Releases hamstring and lower back.

VARIATIONS & ADVANCEMENTS
Use of yoga strap to continually bring the upper body towards the lower body.

BENEFITS
Strengthens the thighs and hamstrings while providing an active recovery from challenging poses.

ENERGETICS
An active relief between chair poses. With continued effort, an empowering pose promoting flexibility.

CUES
Toes point at 12 o'clock. Inhale to lengthen the spine, exhale and sink deeper into the pose. Neck is relaxed and reaches towards the mat. Elbows reach out wide.

ASSISTING
Facing the student from the side, medium pressure can be applied on the low back, and option for the opposite hand to apply pressure to the spine down to the base of the neck.

SANSKRIT TRANSLATION & STORIES
The name (pod-A-hahs-TAHS-anna) comes from the Sanskrit *pada* "foot", *hasta* "hand", and *asana* "posture". Hand Under Foot Pose. There you go!

crow
(BAKASANA)

GENERAL FORM
Flexing forward, knees balance on plank arms as toes touch.

COMMON MISALIGNMENTS
Neck down or craned, core not engaged, knees dumping into elbows, stance too wide.

CONTRAINDICATIONS & MODIFICATIONS
Carpal tunnel syndrome or pregnancy. Modify by keeping feet at "crow prep" or experiment with one foot at time. A block can also be used.

BIOMECHANICS
Connects the upper and lower extremities. Shoulder blades draw forward, stretching both the middle trapezius and the rhomboids.

VARIATIONS & ADVANCEMENTS
Side Crow, Crow to Headstand or Handstand.

BENEFITS
Improves balance and stability. Strengthens wrists and abdominals.

ENERGETICS
Fire, core strength, accomplishment. Zeal!

CUES
Keep your drishti six inches past your mat as you tilt forward. Core is engaged and feet are pressed together.

ASSISTING
No assist - be supportive!

SANSKRIT TRANSLATION & STORIES
Bakasansa (buh-CAW-sun-uh) comes from the Sanskrit *baka* "crane" or *kaka* "crow", and *asana* "posture". While different yoga lineages use one name or another, Dharma Mittra makes a distinction, citing Kakasana as being with arms bent (like the shorter legs of a crow) and Bakasana with arms straight (like the longer legs of a crane). Practitioners in the west often mistranslate the Sanskrit "Bakasana" as "Crow Pose".

equanimity

EAGLE (R,L,R,L)

STANDING LEG RAISE

STANDING LEG RAISE, SIDE

BRING BACK TO CENTER

AIRPLANE

HALF MOON

REPEAT OTHER SIDE

DANCER (R,L,R,L)

TREE (R,L)

MOUNTAIN : FORWARD FOLD : HALFWAY LIFT

VINYASA TO DOWNWARD FACING DOG

eagle
(GARUDASANA)

General Form
Balancing on one leg, squatting, the free leg binds over the standing leg. Arms bind together in front of the face, with fingertips pointing towards the sky.

Common Misalignments
Hunching forward, shoulders up and around ears, shoulders not above hips or hips not square.

Contraindications & Modifications
Arm, hip, or knee injury. Free leg can be placed to a block or to the outside of the standing leg. Arms can come forward for "genie" arms.

Biomechanics
Eagle Pose strengthens the thighs, ankles, and calves while stretching the upper back, shoulders, hips, thighs, ankles and calves. Garudasana also improves coordination, concentration and one's sense of balance. As this pose opens the back lungs, it can increase one's breathing capacity.

Variations & Advancements
Changing up the arm bind by taking genie arms or reverse namaste arms. Touch elbows to knees.

Benefits
Helps with low backache and strengthens ankles and calves.
Additionally, the balance required in Eagle Pose can help protect against knee injuries.

Energetics
Promotes focus and one of the first poses to truly incorporate drishti. Offers a sense of balance, stability, endurance, and strength.

Cues
From Tadasana, bend the knees slightly and lift your right foot to cross your thigh over the left. Point your right toes toward the floor and hook the top of your foot behind your lower left calf. Balance on your left foot. Spread your scapula wide and cross your right arm below your left with elbows bent. Place your left elbow into the crook of your right elbow and raise your forearms perpendicular to the floor. Pressing your palms together, lift your elbows to shoulder height and stretch fingers toward the ceiling. Set your drishti and pull everything into the midline of your body.

Assisting
Medium touch on the hips to help align, outside of the shoulders to encourage pulling into the midline, on the sides of the neck to pull shoulders away from the ears.

SANSKRIT TRANSLATION & STORIES

Garudastana (gar-OOH-duh-STAH-sun-uh) has the root word of Garuda, a legendary bird or bird-like creature from Hindu and Buddhist mythology. Garuda is described as the king of birds, an ever-watchful protector with the power to swiftly go anywhere.

standing leg raise
(UTTHITA HASTA PADANGUSTHASANA A)

GENERAL FORM
Balancing on one leg, the free leg is raised with hands either at hips or holding big toe.

COMMON MISALIGNMENTS
Not square to the top of the mat, torso collapsing, shoulders hunched, raised leg not coming out of hip.

CONTRAINDICATIONS & MODIFICATIONS
Ankle, lower back, or hamstring injury. This pose can be performed against the wall, using a strap to gain access to the toe, or holding the knee.

BIOMECHANICS
Rectus femoris stabilizes the standing leg. The gluteus maximus works with the psoas to balance the pelvis from front to back.

VARIATIONS & ADVANCEMENTS
Grasping the big toe while fully erect in the pose.

BENEFITS
Helps with low backache and strengthens ankles and calves. Stretches muscles between shoulder blades and improves balance.

ENERGETICS
Promotes growth by showing a path to improving the posture, focus, solidarity.

CUES
Firm down into all four corners of your planted foot. Lifted leg comes straight out of hip and toes are engaged to flex forward or point down. Set your drishti and relax your shoulders.

ASSISTING
Medium pressure on the sides of the neck.

SANSKRIT TRANSLATION & STORIES
The name comes from the Sanskrit *utthita* "extended", *hasta* "hand", *pada* "foot", *angustha* "thumb" or "toe", and *asana* "posture". The pose however does not appear to be Indian in origin, and it is not found in the medieval hatha yoga texts.

standing leg raise, side
(UTTHITA HASTA PADANGUSTHASANA B)

GENERAL FORM
Balancing on one leg, the free leg is raised and extended to the side with hands either at hips or holding big toe.

COMMON MISALIGNMENTS
Not square to the top of the mat, torso collapsing, shoulders hunched, raised leg not coming out of hip,

CONTRAINDICATIONS & MODIFICATIONS
Ankle, lower back, or hamstring injury. This pose can be performed against the wall or one the floor, using a strap to gain access to the toe, or holding the knee.

BIOMECHANICS
Abducting and externally rotating the lifted leg. Stretches the hamstring.

VARIATIONS & ADVANCEMENTS
Grasping the big toe while fully erect in the pose. Opposite arm from lifted leg is extended the other direction with gaze following.

BENEFITS
Opens and strengthens hip and pelvic muscles, improves sense of balance.

ENERGETICS
Promotes growth by showing a path to improving the posture, focus, solidarity.

CUES
Continuing to stay strong in your standing leg, open your knee to the side and set your gaze over your opposite shoulder. Maintain spinal integrity and work on centering and opening the hip.

ASSISTING
Medium pressure on the sides of the neck.

SANSKRIT TRANSLATION & STORIES
The name comes from the Sanskrit *utthita* "extended", *hasta* "hand", *pada* "foot", *angustha* "thumb" or "toe", and *asana* "posture". The pose however does not appear to be Indian in origin, and it is not found in the medieval hatha yoga texts.

airplane
(DEKASANA)

General Form
Balanced on one leg, the lifted leg is extended straight back, both arms are extended by the sides, palms facing down with torso facing forward.

Common Misalignments
Rounding the back or craning the neck, standing leg locking up, rotating pelvis, hip of lifted leg rolling up, feet and hands not engaged.

Contraindications & Modifications
Lower back or ankle injuries.
Modify with hands on hips, blocks, or in prayer position.

Biomechanics
Spine extended and supported by erector spinae, quadratus lumborum, and latissimus dorsi.
Rectus abdominis isometrically contracts to fixate the spine.

Variations & Advancements
Hands clasped behind the back, or hands extended forward for Warrior III.

Benefits
Strengthens and tones standing leg, promotes length and improves posture.

Energetics
Strength, stability, lengthening, opening, freedom of expression.

Cues
Keeping a micro bend in your standing leg, toes dial down towards the mat to square the hips. Broaden your chest towards the top of the mat and bring energy to your fingertips. Bring your gaze about two feet in front of your foot, and focus on elongating your body from the crown of your head to your extended heel.

Assisting
Hands come underneath palms to promote deeper opening of the chest and providing stability.

Sanskrit Translation & Stories
Dekasana is a modern pose that is a variation of Warrior III, where Shiva decapitates Daksha with his sword.

half moon
(ARDHA CHANDRASANA)

GENERAL FORM

While balancing on one leg, shoulders, torso, and hips open to side, the free leg is lifted and parallel to the floor, gaze and arms reach towards the sky.

COMMON MISALIGNMENTS

Lifted leg not engaged, hips dropping forward, standing leg locked out or with knee turned out, not spiraling rib cage up.

CONTRAINDICATIONS & MODIFICATIONS

Ankle, lower back or spinal disc injuries. Modify by placing a block under the bottom hand, utilizing a wall for stability, keeping a hand on your hip, or gaze towards the mat.

BIOMECHANICS

The rectus femoris works with the psoas and pectineus to stabilize the leg. Quadriceps straighten the knee and triceps straighten the elbow. The oblique abdominals enable the lower side to bend towards the standing leg.

VARIATIONS & ADVANCEMENTS

Parivrtta Ardha Chandrasana (Revolved Half Moon Pose), has the body revolved towards the standing leg. Baddha Parivrtta Ardha Chandrasana (Bound Revolved Half Moon Pose), has the body revolved towards the standing leg with arms bound around the standing leg.

BENEFITS

Full body stretch, strengthens ankles, thighs, and buttocks. Improves core stability.

ENERGETICS

Freedom of expression, a vulnerable pose full of strength by both rooting down and opening up the heart.

CUES

Ground down through your standing leg, keeping a micro bend in your knee, and lift your opposite leg parallel to the floor, toes flexed towards your face. Ribs spiral towards the ceiling and gaze comes to follow. Use the energy in your top hand to create one long line from fingertip to fingertip.

ASSISTING

Use your hip to stabilize behind the standing leg (like a wall) and use hands to guide lifted arm.

SANSKRIT TRANSLATION & STORIES

The name comes from the Sanskrit *ardha* "half", *chandra* "moon", and *asana* "posture".
Chandra is a lunar deity and is also one of the nine planets in Hinduism. In one myth, Ganesha was returning home on a full moon night after a large feast. A snake crossed his path and frightened Ganesha's mount, dislodging him in the process. An overstuffed Ganesha fell to the ground on his stomach and began to vomit. On observing this, Chandra laughed at Ganesha, who lost his temper. He broke off one of his tusks and flung it straight at the moon, hurting him, and cursed him so that he would never be whole again. Therefore, it is forbidden to behold Chandra on Ganesh Chaturthi. This legend accounts for the moon's waxing and waning, and the large, dark crater on the moon, visible even from earth.

dancer
(NATARAJASANA)

General Form
From Tadasana, shift your weight to your right foot. Bend the left knee and clasp your left hand around the left ankle. Engage your right arm and extend straight out of your shoulder. Set gaze forward.

Common Misalignments
Lifted hip not square to the top of the mat or leg out to side. Standing leg locked out. Lower back compressed. Dumping into the torso by leaning forward.

Contraindications & Modifications
Lower back or shoulder injury. Modify by keeping knee parallel to standing knee or using a strap to create space between the hand and the ankle.

Biomechanics
Lengthening in the psoas and quadriceps of the lifted leg. Muscles in the abdomen and chest loosen to aid in the backbend. Glutes and hamstring pull lifted leg higher with engagement of the latissimus dorsi.

Variations & Advancements
Working towards hand and ankle together at the top of the crown (standing bow).

Benefits
Strengthens spine, thighs, hips, and ankles while stretching thighs, groin, abdominals, shoulder and chest. Improves balance and stamina.

Energetics
Graceful, elegant, rooting strength with opening vulnerability.

Cues
Moving your weight into your right foot, bring your awareness to the four corners of your feet. Extend from your tailbone through the crown of the head and hinge at the hips as you kick your left foot into your left hand. Square your hips and keep your lifted knee in line with your body.

Assisting
Without standing in front of the body, facing the student bring palms to touch to promote lifting the extended arm towards the ceiling, which will open the chest and usually promotes a smile.

SANSKRIT TRANSLATION & STORIES

The name comes from the Sanskrit *nata* "dancer", *raja* "king", and *asana* "posture". Nataraja is one of the names given to the Hindu God Shiva in his form as the cosmic dancer. Nataraja is a well-known sculptural symbol in India and popularly used as a symbol of Indian culture.

tree
(VRKSASANA)

GENERAL FORM
Standing on one leg with opposite foot placed at the inside of the upper thigh. Hands at heart center.

COMMON MISALIGNMENTS
Not in True North alignment, locking out standing leg, pushing rib cage forward.

CONTRAINDICATIONS & MODIFICATIONS
Ankle or knee issues. Raised foot can come to calf, or ankle with toes touching the floor.

BIOMECHANICS
Bones of the upper body are stacked over the long bones of the standing leg. The gluteus maximus in the buttocks works with the psoas to balance the pelvis from front to back. The rectus abdominis tethers the rib cage to the pelvis.

VARIATIONS & ADVANCEMENTS
Extend arms and gaze skyward and/or close eyes.

BENEFITS
Improves posture and promotes body awareness. Brings focus back to breath and improves mental focus.

ENERGETICS
Much like the name suggests, the pose is grounded with room to grow, expand, and open the heart.

CUES
Focus on bringing energy to the center of the body and down, root down through your standing leg. Maintain a neutral pelvis and stretch the sides of your waist upwards towards your fingertips. Imagine one long line from your heels shooting out through the crown of your head.

ASSISTING
Medium pressure on the sides of the neck to allow shoulders to relax.

SANSKRIT TRANSLATION & STORIES
Vrksasana comes from the Sanskrit *vrksa* "tree" and *asana* "posture". A 7th-century stone carving in Mahabalipuram appears to contain a figure standing on one leg, indicating that a pose similar to vrksasana was in use at that time. It is said that Hindu monks disciplined themselves by choosing to meditate in the pose.

grounding

WARRIOR TWO
TRIANGLE
STRADDLE FOLD
PYRAMID
TWISTING TRIANGLE
DOWNWARD FACING DOG
REPEAT OTHER SIDE

triangle
(TRIKONASANA)

General Form
From Warrior II, straighten the front leg and tick-tock the arms to come to the mat and sky.

Common Misalignments
Back rounded instead of rib cage spiraling up, resting into floor or block, knees locked out, heels not in one line, bending from waist instead of hip joint.

Contraindications & Modifications
Neck or lower back injuries, groin strain. Modify by bringing hand to shin or block, shorten stance, bend knees.

Biomechanics
Quadriceps straighten the knees while triceps straighten the elbows. The back-leg buttock muscle extends the leg behind the body and turns it outward. Broadening through chest.

Variations & Advancements
Bring lower hand towards the head (beach ball arms), hand comes to the ground or takes a bind.

Benefits
Opens the shoulders and chest while strengthening muscles around the knees, ankles, hips, and groin.

Energetics
Grounding, an invitation to slow down and focus on the strength of the body before going into backbends.

Cues
Ground down in both feet as you allow your inner thighs to spiral away from each other. Pull your top hip away from the floor and stretch your crown forward, broadening through the chest. Rotate your heart towards the sky and radiate through both arms.

Assisting
Align foot with back foot and guide the top arm skyward.

Sanskrit Translation & Stories
The name comes from the Sanskrit *trikona* "triangle" and *asana* "posture".

straddle fold
(PRASARITA PADOTTANASANA A)

GENERAL FORM
Feet are wide and parallel to each other with the upper torso folded down.

COMMON MISALIGNMENTS
Legs too wide or narrow, hyperextending legs, rounding back.

CONTRAINDICATIONS & MODIFICATIONS
Lower back issues. Modify by using blocks, bending knees, or adjusting width of the stance.

BIOMECHANICS
Both sides of the body are activated and stretched equally. The hips are bent by the psoas, pectineus, and rectus femoris muscles at the front of the thigh. Feet pressed into the mat aid in stability and draw the weight forward.

VARIATIONS & ADVANCEMENTS
Hands can be bound behind the back, take a twist, or advance to tripod handstand.

BENEFITS
Stretches and strengthens hamstrings, calf muscles, groin, and spine.

ENERGETICS
Invites a moment of reflection, pause and strength, and an inversion of energy flow.

CUES
Exhale and hinge forward from your hips. Place your hands on the mat, shoulder width apart. Allow the crown of your head to relax and draw closer to the mat. Invite a bend in your knees.

ASSISTING
Facing the side body, one hand applies light pressure to the sacrum, while the other hand traces the spine.

SANSKRIT TRANSLATION & STORIES
The name comes from the Sanskrit *prasarita* meaning "spread out", *pada* "foot", *uttan* "extended", and *asana* "posture". The pose is not found in medieval hatha yoga texts.

pyramid
(PARSVOTTANASANA)

GENERAL FORM
Standing with one foot in front of the other on "railroad tracks", the front foot points forward and the back foot is slightly turned out with the torso folded over the front leg.

COMMON MISALIGNMENTS
Spine rounded, hips not in line, neck craned.

CONTRAINDICATIONS & MODIFICATIONS
Back, spine, or ankle injuries. Modify by placing hands on hips or blocks, shortening stance or bending knees.

BIOMECHANICS
Both hamstrings are stretched, with the forward fold lengthening the gluteus muscles.

VARIATIONS & ADVANCEMENTS
Taking a bind with the hands,

BENEFITS
Stretches the lower body and strengthens the spine and legs.

ENERGETICS
Continual sense of focus and determination to stay aligned.

CUES
Ground through your feet and square your hips. Keep your back flat as your crown reaches forward, elongating through the spine.

ASSISTING
Hands to hips to encourage alignment. Offer blocks.

SANSKRIT TRANSLATION & STORIES
The name comes from the Sanskrit *parshva* "side", *ut* "intense", *tan* "extend", and *asana* meaning "posture". The pose is not found in medieval hatha yoga texts.

twisting triangle
(PARIVRTTA TRIKONASANA)

GENERAL FORM
Standing with one foot in front of the other on "railroad tracks", the front foot points forward and the back foot is slightly turned out. Torso is folded over the front leg and twisted, the opposite hand comes to the floor and the other to the sky.

COMMON MISALIGNMENTS
Twisting through the spine instead of the shoulder, rounded back, hips not level and parallel to the floor.

CONTRAINDICATIONS & MODIFICATIONS
Lower back or neck injuries. Modify by using blocks or placing hand to the lower back.

BIOMECHANICS
Opens the shoulders and chest while strengthening muscles around the knees, ankles, hips, and groin. Hip flexors allow the pelvis to tip forward.

VARIATIONS & ADVANCEMENTS
Bind arms.

BENEFITS
Rinses toxins in the lower body organs and opens the chest and shoulders. Helps with back pain and stability.

ENERGETICS
A pose of continual growth and focus by staying present, aligned, and balanced.

CUES
Square your hips to the top of the mat and engage your legs, scrubbing your feet and pulling your inner thighs together. Drop your shoulder blades away from your ears and line your arms along a vertical plane.

ASSISTING
Continue to focus on aligned hips, or guide the top hand upward.

SANSKRIT TRANSLATION & STORIES
Para comes from the Sanskrit *parsva* "side". *Vrtta* refers to a unit of limited measure, as in "revolves to a certain point" vs. spinning or continuously linear, *trikona* "triangle", and *asana* "posture".

igniting

HIGH PLANK
LOWER TO THE FLOOR
LOCUST
FLOOR BOW
UPWARD FACING DOG
DOWNWARD FACING DOG
CAMEL
HERO
BRIDGE
WHEEL
SUPTA BADDHA KONASANA
HAPPY BABY

locust
(SALABHASANA)

GENERAL FORM
Lying on the stomach, arms are at sides with shoulders, chest and legs lifted off the floor.

COMMON MISALIGNMENTS
Neck crunched, shoulders up towards ears, legs wide apart.

CONTRAINDICATIONS & MODIFICATIONS
Back, neck or spinal problems/injury. Pregnancy. Modify by placing palms of hands on floor at ribcage (Cobra), forehead on a block, or raise one leg at a time.

BIOMECHANICS
Gluteus maximus extends hips and tilts pelvis downward. Hamstrings extend hips out and up to lift knees. Pectoralis muscles open the chest. Trapezius across back draws shoulders back and down.

VARIATIONS AND ADVANCEMENTS
Small space between legs. Goal post arms. Arms extend forward (Superman). Interlace hands behind back.

BENEFITS
Stretches hip flexors, abdominal organs, thighs, chest, and shoulders. Strengthens the spine, arms, legs, and buttocks. Improves posture. Relives stress. Stimulates abdominal organs. Prepares the body for upcoming backbends in the JIP sequence.

ENERGETICS
Feeling of openness as the body rises up. Willpower. Optimism.
Energy shoots out crown of head to tips of fingers/toes. New awareness and perception arise.

CUES
Gaze straight ahead with your head down and neck neutral. Release your houlders back and down and soften your face. Engage your core and squeeze your bottom. Legs should be hip width or closer. Squeeze legs together and inner thighs towards ceiling. Feet are active reaching towards the back wall. Squeeze arms to side or interlace your fingers behind your back.

ASSISTING
Paddle hands to shoulders. Hold feet.

SANSKRIT TRANSLATION & STORIES
The name comes from the Sanskrit *shalabh,* "grasshopper" or "locust" and *asana,* "posture".
Named after the animal we resemble while in the pose.

floor bow
(DHANURASANA)

GENERAL FORM
Lying on stomach with knees bent, toes flexed back towards the mat. Arms reach behind to grasp ankles. Chest and thighs lift off the mat.

COMMON MISALIGNMENTS
Knees too far apart, feet not active.

CONTRAINDICATIONS & MODIFICATIONS
Knee, shoulder, neck or spinal injury. High or low blood pressure. Headaches. Insomnia. Pregnancy. Modify by using a strap around feet or rolled blanket under thighs to access feet, lift one leg at a time, or look down.

BIOMECHANICS
Posterior deltoids at back of shoulders and triceps extend elbows so hands may grasp ankles. Hamstrings bend knees bringing ankles to hands. Shoulders drawn back and down to open chest.

VARIATIONS AND ADVANCEMENTS
Thighs, calves, and inner feet touch.

BENEFITS
Stretches entire front of body - throat, chest, abdominal muscles and hip flexors, groin, thighs, and ankles. Strengthens back muscles. Improves posture. Stimulates abdominal organs.

ENERGETICS
Body becomes electrified. Feeling of openness. New awareness and perception can arise.

CUES
Engage core muscles and press tailbone down toward the floor to protect your lower back. Gaze is straight ahead. Squeeze your shoulder blades together to open your chest. Flex feet and press them up towards ceiling. Maintain strong legs less than hip width apart. Kick into your hands.

ASSISTING
Hands on feet. Two fingers in mid-back.

SANSKRIT TRANSLATION & STORIES
The name comes from the Sanskrit *dhanura* "bow", and *asana*, "posture".

camel
(USTRASANA)

GENERAL FORM
Kneeling with legs hip width apart and perpendicular to mat. The back is arched with hands reaching towards feet, head is dropped back.

COMMON MISALIGNMENTS
Hips too far forward/back. Core not engaged. Compressing lower back. Straining neck. Legs wider than hip width.

CONTRAINDICATIONS & MODIFICATIONS
Spinal or knee injuries, high or low blood pressure, headache, or vertigo. Modify by placing block between knees/feet, toes turned under with heels elevated. Hands/fists at lower back. Goal post/cactus arms.

BIOMECHANICS
Rhomboids draw shoulders back and down. Pectoralis minor lifts rib cage. Gluteus maximus and hamstrings straighten hips. Quadriceps partially straighten the knees to bring thigh bones to a right angle with floor.

VARIATIONS AND ADVANCEMENTS
Thighs, calves, and inner feet touching.

BENEFITS
Stretches entire front of body - throat, chest, abdominal muscles, and hip flexors, groin, thighs and ankles. Strengthens back muscles. Improves posture. Stimulates abdominal organs. Relieves anxiety and fatigue.

ENERGETICS
Energy and lightness throughout the body. Heart and throat opener. Vulnerability and strength.

CUES
Press shins, ankles, and feet down into the floor. Hips forward and perpendicular to the mat. Tighten core and seat muscles to protect lower back. Squeeze shoulder blades together and down. Lift chest up towards sky. Release neck. Continue to gain length on your inhale and open your chest on the exhale.

ASSISTING
Two fingers to the back point of the thoracic spine, approximately at the bra line.

SANSKRIT TRANSLATION & STORIES
The name comes from the Sanskrit *ustra* "camel", and *asana,* "posture". Nothing official, but most likely named as such because the intense backbend resembles a camel's hump, or the way that camels bend their knees and fold their legs beneath their bodies to lay down.

hero
(VIRASANA)

General Form
Kneeling tall on mat with knees together, sitting on feet. Palms are on thighs and gaze is forward.

Common Misalignments
Shoulders by ears. Rushing through or skipping pose.

Contraindications & Modifications
Knee or ankle injury. Heart problems. Headache.
Modify by placing blocks or folded blanket beneath hips. Eyes closed.

Biomechanics
Knees are bent and weight is set back on the heels, constricting blood flow to the backs of the knees while in the pose. Lower belly pulled in to lengthen the lower back. Shoulders are integrated and chest is open.

Variations and Advancements
Feet wider than hips with buttocks on floor, recline backwards to rest on elbows or the floor (Reclining Hero).

Benefits
Stretches thighs, hips, and ankles. Reduces swelling of legs. Improves digestion and menopause symptoms.
Alleviates high blood pressure. Calms the mind.

Energetics
Calming, strong, steady. Grounding and emotionally soothing.

Cues
Relax your shoulders and let your hands rest comfortably on your lap. Reach your crown upwards.
Focus on one point ahead or close your eyes. Breathe.

Assisting
Hands to shoulders encouraging student to relax shoulders down.
Block if knees out wide or if student looks uncomfortable.

Sanskrit Translation & Stories
The name comes from the Sanskrit *vira* "hero", and *asana,* "posture".
The root word *vir* means "to overpower".

bridge
(SETU BANDHA SARVANGASANA)

GENERAL FORM
From your back, shoulders and feet stay on mat while hips lift.

COMMON MISALIGNMENTS
Pressing into heels and lifting toes, knees too wide, chin tucked into chest.

CONTRAINDICATIONS & MODIFICATIONS
Back, neck, or shoulder injury. High blood pressure. Modify by using bolsters or block underneath the lower back or adjusting placement of heels.

BIOMECHANICS
Glutes stabilize the spine while rhomboids and hamstrings stretch to lift the frontal plane. The entire front body stretches including quadriceps, psoas, abdominals, and pectoralis major and minor.

VARIATIONS & ADVANCEMENTS
Alternate lifting legs or raising heels. Bind hands behind back or clasp sides of ankles.

BENEFITS
Strengthens thighs, buttocks, and abdominals while opening the chest. Stimulates digestion and thyroids and improves circulation of blood. Prepares the body for wheel pose.

ENERGETICS
Vigilant, prepared, strong base, sturdy and ready to take on the next pose. Room for growth and expansion.

CUES
Root down through the soles of your feet and contract your quads as you lift your hips. Open your heart and roll your shoulders beneath you. Lengthen your neck away from your shoulders.

ASSISTING
Two fingers to the back point of the thoracic spine, approximately at the bra line.

SANSKRIT TRANSLATION & STORIES
The pose is named from the Sanskrit *setu*, "bridge" *bandha*, "caught", *sarva*, "all", *anga*, "limb", and *asana*, "posture". It is said the Hindu god Rama built a bridge from the Indian town of Rameswaram across the sea to Sri Lanka to rescue his wife Sita from her abductor Ravana. The Ramanathaswamy Temple, dedicated to the Hindu god Shiva, is at the center of the town and is closely associated with Rama. The temple and the town are considered a holy pilgrimage site, and present day is the closest point from which to reach Sri Lanka from India.

wheel
(CHAKRASANA)

GENERAL FORM
Extended backbend. Hands are next to ears, feet on mat are hip width distance, hip and pelvis are the highest points towards the sky with all four posts pushing and lifting the body.

COMMON MISALIGNMENTS
Knees too wide, crunched neck or compressed lower back. Weight not distributed equally to all four limbs. Elbows splayed out.

CONTRAINDICATIONS & MODIFICATIONS
Shoulder or back injury, carpal tunnel syndrome, headache, high or low blood pressure. Modify by staying in bridge or coming to the crown of the head with arms at right angles. Utilize wall to work up to full wheel.

BIOMECHANICS
The erector spinae and latissimus dorsi contract powerfully to move spine into its extended position. Hip flexors lengthen & gluteus maximus muscles are in concentric contraction. Triceps contract and the deltoids lengthen to let the shoulder joint open.

VARIATIONS & ADVANCEMENTS
Lifting one leg up (lifts out of hip, toes pointed to sky) or come onto toes.

BENEFITS
Increases energy and oxygen flow to the blood. Opens and releases the tightness/tension in the upper back, chest, shoulders, and hip flexors. Energizes the entire body, mentally uplifting.

ENERGETICS
Intimidating yet exhilarating! Feeling of accomplishment, pushing the edge and coming to the paramount. Powerful, childlike, embodies physical freedom, lightness, and joy.

CUES
From bridge, hands come to your ears, press up as you exhale. Keep a micro bend in the elbow. Root into all four limbs and rotate inner thighs towards the floor. Let your neck hang free.

ASSISTING
Two fingers to the back point of the thoracic spine, approximately at the bra line.

SANSKRIT TRANSLATION & STORIES
Known also as Urdhva Dhanurasana, the names come from the Sanskrit *chakra*, "wheel", *urdhva* "upwards", *dhanura* "bow" (for shooting arrows), and *asana* "posture".

supta baddha konasana
(POSE CALLED IN SANSKRIT)

GENERAL FORM
Lying on the back, soles of the feet come together (heels to groin) with knees out wide. Arms rest where comfortable.

COMMON MISALIGNMENTS
Forcing knees onto the floor, allowing lower back to lift.

CONTRAINDICATIONS & MODIFICATIONS
Groin, knee, or hip injury, lower back pain. Modify by taking Fallen Bridge (feet planted wide on mat, knees coming to touch), support thighs on blocks.

BIOMECHANICS
Releases the lower back and hips while stretching the inner thigh muscles.

VARIATIONS & ADVANCEMENTS
Arms come to cactus to open up the chest. Elevate pelvis by using a block.

BENEFITS
Helps relieve symptoms of stress, depression, menstruation, and menopause. Soothes the nervous system.

ENERGETICS
Rest, deep connection, pride of accomplishment.

CUES
Bring your feet together and let your belly drop. Take your arms to cactus or bring one hand to your heart and one to your belly. Close your eyes, breath, and let yourself melt into the mat.

ASSISTING
Hands to shoulders, medium pressure towards the neck or mat.

SANSKRIT TRANSLATION & STORIES
The name comes from the Sanskrit *supta* "reclined", *baddha* "bound", *kona* "angle", and *asana* "posture". Also considered cobblers pose, coming from the position Indian cobblers are seated.

happy baby
(ANANDA BALASANA)

GENERAL FORM
Lying on the back with legs straddled open and upward, legs are bent and toes are flexed back towards the face. Hold soles of the feet with hands.

COMMON MISALIGNMENTS
Back not plugged into the mat, curved spine, tailbone lifted, straining of the neck.

CONTRAINDICATIONS & MODIFICATIONS
Pregnancy, knee or neck injury. Modify by using a blanket for neck support, utilizing a yoga strap around the arches of the feet, or bringing hands to calves.

BIOMECHANICS
Lengthens the spine and stretches hamstrings and inner thighs. Gentle opening of the hip joints.

VARIATIONS & ADVANCEMENTS
Sway side to side, like a happy baby.

BENEFITS
Brings body to a restorative and resting state.

ENERGETICS
Surrender, restoration and relaxation. Childlike freedom.

CUES
From your back, straddle your legs and access your feet, gently pull your feet, knees and quadriceps towards the floor. Open knees wide and roll your tailbone towards the mat. Rock side to side.

SANSKRIT TRANSLATION & STORIES
Ananda Balasana comes from the Sanskrit *ananda* "happy", *bala* "baby", and *asana* "posture".

stability

SCISSOR LEGS

30/60/90/LIFT

ABDOMINAL TWIST

BOAT

VINYASA OR STEP TO DOWNWARD FACING DOG

scissor legs
(POSE CALLED IN ENGLISH)

GENERAL FORM
Lying on your back with hands behind your head, one leg hovers above the mat, with the other raised straight out of the hip. Crunch upwards ten times.

COMMON MISALIGNMENTS
Craning of the neck, feet/abdominals not engaged, lower back lifted off the mat.

CONTRAINDICATIONS & MODIFICATIONS
Lower back or neck injury. Modify by bending the knees or placing the lower foot on the ground.

BIOMECHANICS
Engages the low belly, quadriceps, and calves.

VARIATIONS & ADVANCEMENTS
"Scissor" legs and switch leg placement after each crunch. Use pointer fingers to focus crunch across body to engage obliques.

BENEFITS
Maintains stability and balance and improves core strength.

ENERGETICS
Stability, strength, transformation, and focus.

CUES
Engage through your toes to fire up your leg muscles. Shoulder blades come off of the floor and gaze comes to your toes. Press through the balls of your feet and exhale as you lift.

ASSISTING
Gentle touch to the lifted leg to offer support and stability.

30/60/90 lift
(POSE CALLED IN ENGLISH)

GENERAL FORM
Lying on your back with hands behind under your seat, legs raise straight out of the hip to begin. Legs are engaged as they lower 30 degrees, then 60 degrees, then hover, and return upwards.

COMMON MISALIGNMENTS
Craning of the neck, feet/abdominals not engaged, lower back lifted off the mat.

CONTRAINDICATIONS & MODIFICATIONS
Lower back or neck injury. Modify by hovering one leg at a time or leaving legs upright.

BIOMECHANICS
Engages the entire core, quadriceps, and calves.

VARIATIONS & ADVANCEMENTS
Holding the pose for longer!

BENEFITS
Maintains stability and balance and improves core strength.

ENERGETICS
Stability, strength, transformation, and focus.

CUES
Engage your toes to fire up your leg muscles. Shoulder blades come off of the floor and gaze comes to your toes. Press through the balls of your feet and exhale as you lower.

ASSISTING
Hand follows the feet to offer support and security.

abdominal twist
(POSE CALLED IN ENGLISH)

GENERAL FORM
Lying on your back with hands behind your head, elbows wide, legs are bent with knees above hips and feet engaged. Opposite elbow comes to opposite knee as you crunch through center ten times.

COMMON MISALIGNMENTS
Craning of the neck, feet not engaged, bringing knees past the hips (closer to face).

CONTRAINDICATIONS & MODIFICATIONS
Lower back or neck injury. Modify by keeping legs straight or hold at 90 degrees.

BIOMECHANICS
Engages the obliques and opens the chest.

VARIATIONS & ADVANCEMENTS
Increase repetitions or take the crunches slowly.

BENEFITS
Maintains stability and balance and improves core strength.

ENERGETICS
Stability, strength, transformation, and focus.

CUES
Engage your toes to fire up your leg muscles. Shoulder blades come off of the floor and gaze comes to your knees. Press through the balls of your feet and exhale as you crunch. Elbows stay wide as chest reaches towards the sky and reaches over.

ASSISTING
Gentle touch (like paddy-cake) to the feet to offer support and security.

boat
(NAVASANA)

GENERAL FORM
Sit balanced between the tail bone & sits bones. Legs and arms reach upwards to create a "V".

COMMON MISALIGNMENTS
Rounding of the back or straining neck.

CONTRAINDICATIONS & MODIFICATIONS
Neck and lower back injury, menstruation. Modify by bending the knees, holding the backs of the thigh, bringing heels to the mat, or using a strap,

BIOMECHANICS
Quadriceps engage to lift legs as hip flexors are concentrically shortened, moving the legs toward the torso and the hips into a flexed position.

VARIATIONS & ADVANCEMENTS
Half Boat, taking toes with peace fingers, or alternating from Boat to Low Boat.

BENEFITS
Stimulates the kidneys, thyroid, prostate and intestine. Strengthens the core and lower back.

ENERGETICS
Balance, strength, transformation, fun, challenging, lighthearted, empowering, stability. A safe pose to challenge yourself, where it's fun to "fail" as your core becomes stronger. Usually a humorous point in the yoga class.

CUES
Palms face each other as you reach towards the sky, feet are activated and legs are strong. Relax your neck. Open your sternum to the ceiling. Hold for 5...4...3... (motivational counting).

ASSISTING
Two fingers to the back point of the thoracic spine, approximately at the bra line.

SANSKRIT TRANSLATION & STORIES
The name comes from the Sanskrit *nava* "boat" and *asana* "posture".

opening

HALF PIGEON

DOUBLE PIGEON

REPEAT OTHER SIDE

FROG

half pigeon
(EKA PADA RAJAKAPOTASANA)

GENERAL FORM
Front leg bends toward the opposite wrist with knee and shin on the mat and straightened back leg. Hips are squared towards the mat. Torso folds over legs.

COMMON MISALIGNMENTS
Hips not square, back leg out to the side.

CONTRAINDICATIONS & MODIFICATIONS
Knee or lower back injuries. Modify by placing a yoga block under the front hip or hip closer to the bent knee. Take position on your back, making a figure four with legs.

BIOMECHANICS
Stretches the front leg hip muscles, most notably the glutes and adductor magnus. On the back leg, the hip flexors are lengthened as the hip moves into extension.

VARIATIONS & ADVANCEMENTS
King Pigeon or Double Pigeon.

BENEFITS
Stretches pelvis muscles, improves flexibility and posture.

ENERGETICS
Invites vulnerability, allows entire body to relax as the practice comes to a close.
Gratitude, transformative growth.

CUES
From Downward Facing Dog, bring your front leg to a 90-degree angle while your back leg extends behind you. Square your hips and lengthen your spine as your lean forward. Surrender into the pose.

ASSISTING
Pressure on the low back, guiding hips lower.

SANSKRIT TRANSLATION & STORIES
The name comes from the Sanskrit *eka* "one", *pada* "leg", *raja* "king", *kapota* "pigeon" and *asana* "posture".

double pigeon
(AGNISTAMBHASANA)

GENERAL FORM
Seated with knees bent, one leg is crossed over the other, stacking shins, legs at right angles.
Torso folds over legs.

COMMON MISALIGNMENTS
Collapsing spine, legs to wide, rotating from knees instead of hips.

CONTRAINDICATIONS & MODIFICATIONS
Knee or lower back injuries. Modify by placing blocks under knees. Can also be taken while reclined.

BIOMECHANICS
Engages the transverse abdominis, stretches the psoas, sartorius, iliacus, and other hip and thigh muscles.

VARIATIONS & ADVANCEMENTS
Lower your head and hands to the mat.

BENEFITS
Double Pigeon is an excellent stretch for the hips and groins, and also stretches the glutes and the lower back.
It is therapeutic for stress, and aids with relaxation, calming the mind.

ENERGETICS
Grounding, opening, relaxing, and challenging.

CUES
Keep your right foot flexed and draw your right knee away from you until you feel a stretch in your right hip.
Line up the side of your right foot over your left knee, and your right knee over the side of your left foot. Try
to make your shins parallel to each other, with the right stacked directly over the left. Breathe here, and opt to
walk your hands forwards, surrendering to the mat.

ASSISTING
Grounding pressure on the lower back, grounding to the mat.

SANSKRIT TRANSLATION & STORIES
Agnistambhasana is often called "fire log" pose, and comes from the Sanskrit *agni* "fire", *stampbha* "pillar",
and *asana* "posture". Agni, the Vedic fire god of Hinduism, is noted in classical Indian religious cosmology as
one of the five inert impermanent constituents (*Dhatus*) along with space (*Akasha/Dyaus*), water (*Jal*), air
(*Vayu*) and earth (*Prithvi*), the five combining to form the empirically perceived material existence (*Prakriti*).

frog
(MANDUKASANA)

GENERAL FORM
Lying on the stomach with the knees bent at 90-degrees directly out of the hips, feet flexed.

COMMON MISALIGNMENTS
Hips too far back or forward, collapsing into the lower back.

CONTRAINDICATIONS & MODIFICATIONS
Hip, knee, or lower back injury. Modify by using a wall, blocks to support head or thighs, or place a bolster under the torso.

BIOMECHANICS
Engaging abdominals as stabilizers. Pelvis is at an anterior tilt with the erector spinae engaged.

VARIATIONS & ADVANCEMENTS
Full straddle fold lying on stomach or against the wall.

BENEFITS
Rejuvenating deep stretch, releases muscle tension in the hips and groin.

ENERGETICS
Intense and rejuvenating. Intimidating and full of challenge.
Lots of internal struggle which is relieved the longer the pose is held.

CUES
From table top, walk your knees out wider than your hips. Flex your feet so your toes face outward and your heels are directly behind your knees. Option to place a blanket under your knees. Walk your hands forward. Reach the crown of your head forward and your tailbone back. Keep your hips in the same plane as your knees. Lift your belly away from the ground.

ASSISTING
Ensure students have proper alignment, or bring attention to energy in the feet.

SANSKRIT TRANSLATION & STORIES
The name comes from the Sanskrit *manduka*, "frog" and *asana* "posture".

release

SEATED SINGLE LEG EXTENSION (R,L)
SEATED FORWARD BEND
TABLE TOP/INCLINE PLANE
FISH

seated single leg extension
(JANU SIRSASANA)

GENERAL FORM
Seated with one leg extended, the other leg bends with sole of the foot coming to the inner thigh.
Torso bends over extended leg.

COMMON MISALIGNMENTS
Not sitting tall on sits bones. Bending at the waist instead of hinging at the spine.

CONTRAINDICATIONS & MODIFICATIONS
Knee or lower back injury. Modify by adjusting the bend in the extended leg or use a strap.

BIOMECHANICS
Obliques and rectus abdominus are engaged in the forward fold. Releases tension in the lower back.

VARIATIONS & ADVANCEMENTS
Use a partner. Widening the angle between the legs past 90 degrees. Instead of bringing the bent-knee heel
into the perineum, snug it into the same-side groin.

BENEFITS
Relaxes calves, hamstrings, glutes, and back. Improves blood circulation in the legs, dissolves restlessness and
irritability, soothes the brain and nervous system.

ENERGETICS
Transformative growth, focus, release.

CUES
Inhale, bend your knee and draw the heel back toward your perineum. Rest the sole of your foot against your
inner thigh. Line up your navel with the middle of the extended thigh. With the arms fully extended, lengthen
the front torso from the pubis to the top of the sternum. Exhale and extend forward from the groin, not the
hips. As you descend, bend your elbows out to the sides and lift them away from the floor.

ASSISTING
Pressure on the lower back, grounding towards the mat.

SANSKRIT TRANSLATION & STORIES
The name comes from the Sanskrit *janu* "knee", *shirsha* "head", and *asana* "posture".

seated forward bend
(PASCHIMOTTANASANA)

GENERAL FORM
Seated with both legs extended, torso bends forward.

COMMON MISALIGNMENTS
Not sitting tall on sits bones. Bending at the waist instead of hinging at the spine.

CONTRAINDICATIONS & MODIFICATIONS
Knee or lower back injury. Modify by adjusting the bend in the legs or use a strap.

BIOMECHANICS
Stretches semitendinosus, bicep femoris, semimembranosus, gastrocnemius. Releases tension in the lower back.

VARIATIONS & ADVANCEMENTS
Use a partner or assisted forward bend.

BENEFITS
Relaxes calves, hamstrings, glutes, and back. Improves blood circulation in the legs, dissolves restlessness and irritability, soothes the brain and nervous system.

ENERGETICS
Transformative growth, focus, release.

CUES
Inhale as you extend your arms upward, lengthen the front torso, exhale and extend forward out of the hips. As you descend, bend your elbows out to the sides and lift them away from the floor.

ASSISTING
Pressure on the lower back, grounding towards the mat.

SANSKRIT TRANSLATION & STORIES
The name comes from the Sanskrit *paschima* "west" or "back of body", *uttana* "intense stretch", and *asana* "posture".

table top
(ARDHA PURVOTTANASANA)

GENERAL FORM
Feet are flat on the floor, hip width distance with knees bent at 90 degrees, pushing hips into a plank position. Palms are placed on floor with fingers facing toes, torso faced upwards.

COMMON MISALIGNMENTS
Knees not aligned over ankles, feet too wide, sagging seat, shoulders not aligned over wrists, neck compressed.

CONTRAINDICATIONS & MODIFICATIONS
Wrist, lower back, or neck injury. Modify by practicing the pose in a chair.

BIOMECHANICS
Stretches the pectoralis major and minor, and anterior deltoid. Quadriceps engage for stability. Triceps straighten the elbows, lengthening the biceps.

VARIATIONS & ADVANCEMENTS
Alternate lifting legs.

BENEFITS
Strengthens the upper body and opens the chest. Neutralizes forward bends.

ENERGETICS
Strength, opening, release. Freedom.

CUES
From seated bend your knees and plant your hands under your shoulders, fingers facing toes. Exhale as you press up. Engage your thighs and hamstrings. Let the head hang.

ASSISTING
Two fingers to the back point of the thoracic spine, approximately at the bra line.

SANSKRIT TRANSLATION & STORIES
The name, also known by Half Upward Plank pose, Half Reverse Plank, or Crab, comes from the Sanskrit *ardha* "half", *purva* "east", *uttana* "intense stretch", and *asana* "posture".

incline plane
(PURVOTTANASANA)

GENERAL FORM
With torso faced upwards, legs are extended straight and arms are placed on floor with fingers facing toes. Hips are raised towards the sky and head hangs towards the floor.

COMMON MISALIGNMENTS
Hips sinking or legs not engaged. Shoulders not stacked over hips. Compressed neck.

CONTRAINDICATIONS & MODIFICATIONS
Wrist or neck injury. Modify by leaving neck neutral, or practicing using a chair. Tabletop pose.

BIOMECHANICS
The back is arched by the erector spinae along the spine and the quadratus lumborum in the lower back. The posterior deltoid muscles extend the shoulders back and away from the torso.

VARIATIONS & ADVANCEMENTS
Alternating lifting legs.

BENEFITS
Neutralizes forward bends by releasing the back and opening the front of the torso. Tones the shoulders and biceps, strengthens the arms, wrists, and legs.

ENERGETICS
Release, expansion stretching the entire body, length, openness.

CUES
From seated, straighten legs out in front of you, placing your hands underneath your shoulders with fingers facing toes. Press down into your hands to lift your hips and stack shoulders over wrists. Move your shoulder blades in towards each other. Engage your thighs and bring your drishti to the wall behind you.

ASSISTING
Two fingers to the back point of the thoracic spine, approximately at the bra line.

SANSKRIT TRANSLATION & STORIES
The name comes from the Sanskrit *purva* "east", *uttana* "intense stretch", and *asana* "posture".

fish
(MATSYASANA)

General Form
Lying on the back, elbows prop under the upper back to lift and arch torso up, head is released back.

Common Misalignments
Hips lifted, weight into head and neck, elbows and forearms not pulled in towards torso.

Contraindications & Modifications
Back or neck injury, headache, high or low blood pressure.
Modify by using bolsters or blocks under the low back.

Biomechanics
Engages rhomboideus and trapezius to lift upper torso. Opens and stretches the front side of the body.

Variations & Advancements
Knees bent, feet on floor. Legs bound together. Toes can be lifted about six inches.
Hands placed to heart in Anjali Mudra.

Benefits
Opens the throat, chest, and torso, redirects blood flow through the thyroid and parathyroid glands.

Energetics
Heart opening, restorative, supported and balanced.

Cues
From the mat, take your arms by your sides and push down into your heels to left your hips. Place your hands beneath your seat, palms facing down. Elongate your legs and press down into your forearms, bending your elbows. Lift with your chest to bring your head off of the floor. Create an arch in your back and gently tilt your head, bringing it back to the mat.

Assisting
Two fingers to the back point of the thoracic spine, approximately at the bra line. Hold onto the feet.

Sanskrit Translation & Stories
The name comes from the Sanskrit *matsya* "fish" and *asana* "posture".

rejuvenation

SHOULDER STAND
PLOW
EAR PRESSURE

OR HEADSTAND
OR HANDSTAND
CHILD'S POSE

OR LEGS UP THE WALL

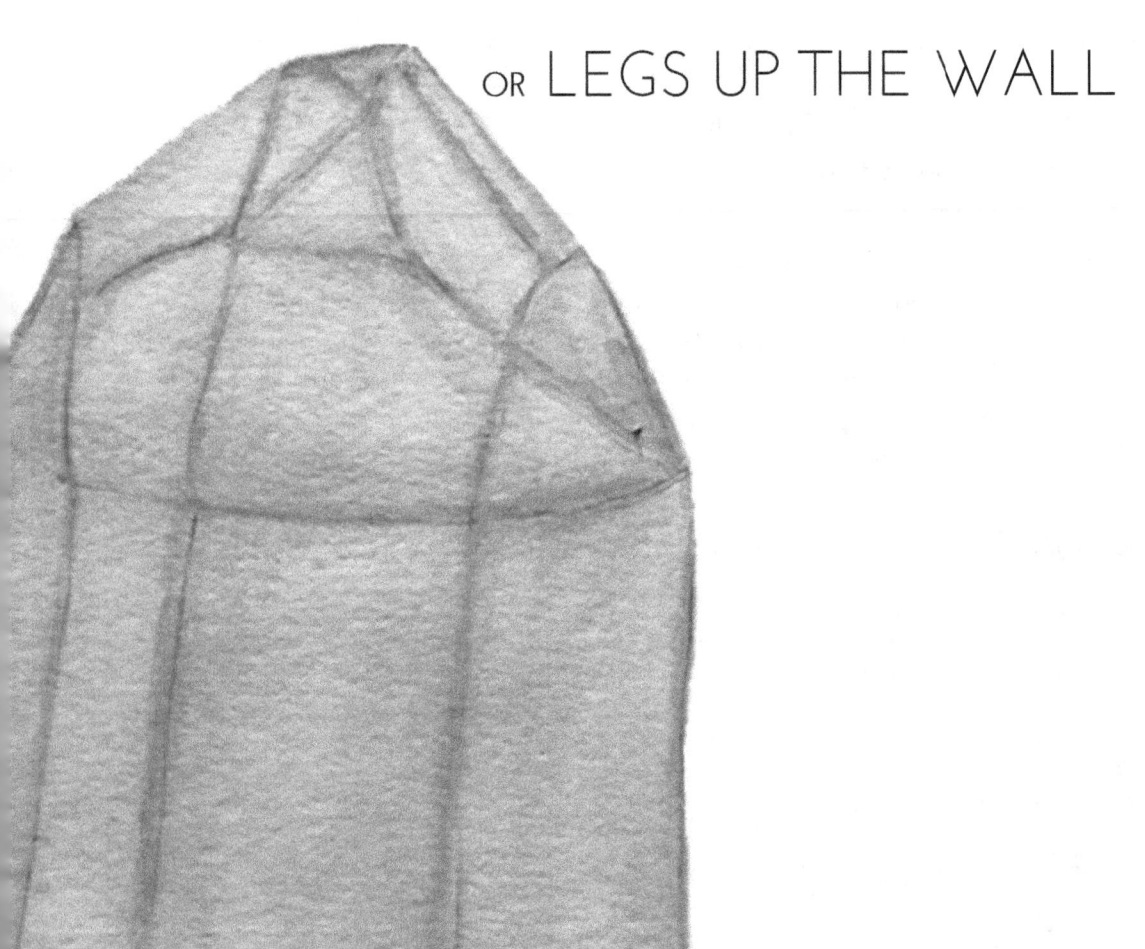

shoulder stand
(SALAMBA SARVANGASANA)

GENERAL FORM
Lying on the back with weight in shoulders, legs are straight up with hands supporting the low back and chin tucked into chest.

COMMON MISALIGNMENTS
Strain in the neck, hips resting in hands, legs not engaged, elbows splayed out.

CONTRAINDICATIONS & MODIFICATIONS
High blood pressure, neck or back issues, headache. Modify by using a wall or place a block under tail bone.

BIOMECHANICS
Shoulder joint is extended and the chest opens. The erector spinae and the rectus abdominus lift the trunk. The pelvis is supported and held level by the gluteus maximus in the buttocks and the psoas high at the front of the thighs and in the pelvis.

VARIATIONS & ADVANCEMENTS
Legs in Locust or Supta Konanasa, toes on the ground.

BENEFITS
Improves respiratory function. Strengthens the muscles of the back and abdomen. Drains legs of metabolic waste and allows fresh, oxygen rich blood to circulate through the head, heart, and chest.

ENERGETICS
Playful, rejuvenating. Balance and inner peace.

CUES
From your back, bring legs and hips to the sky, rolling back onto your shoulders. Hands come to your lower back and walk shoulder blades and elbows in towards each other. Engage your legs by flexing or pointing your toes. Open your chest.

ASSISTING
Bring attention to the feet to have practitioner engage.

SANSKRIT TRANSLATION & STORIES
The name comes from the Sanskrit *salamba*, "supported", *sarvanga*, "all limbs/the whole body", and *asana*, "posture"

plow
(HALASANA)

GENERAL FORM
Lying on the back with weight in shoulders, the upper torso is lifted straight up, lower body flexed back over the head with toes touching on the floor behind the head. Hands come to the low back.

COMMON MISALIGNMENTS
Head turned, rounding of the spine, neck not relaxed. Rushing through the pose.

CONTRAINDICATIONS & MODIFICATIONS
Neck or lower back injuries, menstruation or pregnancy. Modify by using a folded blanket under shoulders, using a wall, bending knees, or bringing a block behind your head.

BIOMECHANICS
The biceps flex the elbows, causing the hands to press into the back, lifting and supporting it while opening the chest. The quadriceps extend the knees, triggering reciprocal inhibition, relaxing the hamstrings.

VARIATIONS & ADVANCEMENTS
Supta Konasana

BENEFITS
Beneficial effects on the cardiovascular system and the flow of cerebral spinal fluid. This fluid bathes the brain in endorphins and improves circulation to regions of stagnant flow.

ENERGETICS
Playful, rewarding, challenging, soothing.

CUES
From shoulder stand, use your core to slowly lower your legs to the floor behind you. Spin your inner thighs up to the ceiling and press through your tailbone. Gaze at your navel and relax your neck.

ASSISTING
Look to grab blocks

SANSKRIT TRANSLATION & STORIES
The name comes from the Sanskrit *hala* "plow" and *asana*, "posture". The pose is described and illustrated in the 19th century *Sritattvanidhi* as Langalasana, another word for plow in Sanskrit.

ear pressure
(KARNAPIDASANA)

GENERAL FORM
Lying on the back with weight in shoulders, the upper torso is lifted straight up, lower body flexed back over the head, knees are bent and resting beside ears.

COMMON MISALIGNMENTS
Legs forced down, crunching into neck.

CONTRAINDICATIONS & MODIFICATIONS
Neck or back injuries, menstruation, pregnancy.
Modify by using a folded blanket under shoulders or a block behind your head.

BIOMECHANICS
Major muscle groups are stretched, improving flexibility of the spine, shoulders, and legs. The spinal extensor muscles are lengthened and the muscles of the neck are deeply stretched.

VARIATIONS & ADVANCEMENTS
Shins and forearms flush on the mat, or equally at 45 degrees away from the mat, with intention.

BENEFITS
The overarching spinal column stretch gives a deep massage to the internal organs, especially the intestine. The stomach fold stimulates digestion to increase metabolic rate.

ENERGETICS
Playful, rewarding, challenging, soothing.

CUES
From Plow, bring your knees in towards your ears.
Sweep your arms around to take a bind, pressing into the mat.

SANSKRIT TRANSLATION & STORIES
The name comes from the Sanskrit *karna* "ears" and *asana*, "posture". Also known as Deaf Man's Pose.

headstand
(SIRSASANA)

GENERAL FORM
Upside down, head on the mat, cradled in hands, forearms parallel to floor with body straight in the air.

COMMON MISALIGNMENTS
Weight not drawn upwards; body is in slouched position. Elbows out too wide.

CONTRAINDICATIONS & MODIFICATIONS
Back or shoulder injuries, headache, high blood pressure or cardiovascular issues, pregnancy. Modify by resting your hand clasp, shoulders, hips, and upturned heels with the support of a wall.

BIOMECHANICS
The adductor group draws the thighs together as the quadriceps straightens the knee. Muscles running the length of the spine lift the back into the pose and remain active to stabilize it. The lower trapezius draws the shoulders away from the neck, freeing the cervical spine.

VARIATIONS & ADVANCEMENTS
Eagle, lotus, or splits with legs. Pike up and down for abdominal work.

BENEFITS
Inverting the body stimulates the control mechanisms in the heart and the arteries that monitor and adjust blood pressure. Positively affects the flow of cerebrospinal fluid in the spinal cord and the brain.

ENERGETICS
Rejuvenating, intimidating, challenging, playful.

CUES
Forearms are parallel to each other, aligned with your shoulders. Tuck your tailbone in and push through your forearms, head and chest lifting away from the flow. Draw your legs and torso up towards the ceiling and visualize your body as one long line of energy.

SANSKRIT TRANSLATION & STORIES
Also known by the name *Salamba Sirsasana,* the pose comes from the Sanskrit s*alamba* "supported", *sirsa* "head", and *asana* "posture". In the 11th century the pose was called *Duryodhanasana (dur* "extremely hard", *yodhana* "fight;" - "the one with whom the fight is extremely hard".
Also seen in the 18th century as *kapalikarana* "skull technique".

handstand
(ADHO MUKHA VRKSASANA)

GENERAL FORM
The body is supported in a stable, inverted vertical position by balancing on the hands. The body is held straight with arms and legs fully extended, hands spaced approximately shoulder-width apart and the legs together.

COMMON MISALIGNMENTS
Not aligned in True North, swaying, craning neck, sinking into shoulders.

CONTRAINDICATIONS & MODIFICATIONS
Shoulder or back injury, carpal tunnel syndrome, headache.
Modify by using a wall to support back or feet, or use handstand hops.

BIOMECHANICS
The adductor group draws the thighs together as the quadriceps straightens the knee. Muscles running the length of the spine lift the back into the pose and remain active to stabilize it. The lower trapezius draws the shoulders away from the neck, freeing the cervical spine.

VARIATIONS & ADVANCEMENTS
Eagle, lotus, or splits with legs. Pike up and down for abdominal work.
Stag split, in which legs are front split with bent knees.
Back extremely arched, with bent knees and toes touching the back of the head.
Hollow back, with hyperextension of the back so that legs are held further back than the head.
One-handed, in which only one hand contacts the ground.
Handstand pushups, in which one raises and lowers the body while standing inverted on the hands.
Straddle split handstand

BENEFITS
Inverting the body stimulates the control mechanisms in the heart and the arteries that monitor and adjust blood pressure. Positively affect the flow of cerebrospinal fluid in the spinal cord and the brain.

ENERGETICS
Rejuvenating, intimidating, challenging, playful.

CUES
Begin with one foot lifted to the sky and hop to bring the other foot up. Engage your abdominal muscles for stability, lengthen your spine, and move shoulder blades away from your head and neck. Elongate your entire body from your shoulders all the way down into your toes. Gaze is neutral or between your hands.

Sanskrit Translation & Stories
The name comes from the Sanskrit *adho* "downward", *mukha* "facing", *vrksa* "tree", and *asana* "posture".

deep rest

SUPINE TWIST (R,L)
SAVASANA

supine twist
(JATHARA PARIVARTANASANA)

GENERAL FORM
Lying on the back, knees fall to one side of the body, twisting the spine.

COMMON MISALIGNMENTS
Shoulder blades lifting off floor, tensing up in shoulders or ears.

CONTRAINDICATIONS & MODIFICATIONS
Shoulder or lower back injury. Modify by using folded blanket or block under knees.

BIOMECHANICS
Rectus abdominus is engaged, spine stretches.

VARIATIONS & ADVANCEMENTS
Legs can be straight, twist one knee at a time, turn head. Use hands to guide knee/leg towards the mat.

BENEFITS
Flushes spine with nutrient rich blood, loosens hips, tones abdominals.

ENERGETICS
Cleansing, release, stretching, completion, rest.

CUES
Stack your hips and release your legs to one side. While keeping your shoulder blades rooted to the mat, take your gaze over your opposite shoulder. Breathe into your side body.

SANSKRIT TRANSLATION & STORIES
The name comes from the Sanskrit *jathara*, "stomach" or "abdomen", *parivartana* "to turn around", and *asana* "posture".

savasana
(POSE CALLED IN SANSKRIT)

GENERAL FORM
Lying on the back with legs spread as wide as the yoga mat and arms relaxed to the side.
The eyes are closed and the breath is deep.

COMMON MISALIGNMENTS
Tensing muscles, neck not aligned with spine.

CONTRAINDICATIONS & MODIFICATIONS
Back injury. Modify by using blocks or blankets to find most comfortable position.

BIOMECHANICS
Body is in full resting position, neutral alignment, restoring all systems.

VARIATIONS & ADVANCEMENTS
Feet flat on floor, palms facing up.

BENEFITS
Deep rest, inviting in peace and spiritual awakening. Rejuvenates the body, mind, and spirit.

ENERGETICS
Calm, peace, healing, restoration, open, aligned, completion.

CUES
Come to your back and allow your body to fully rest. Palms face the sky to receive, or the mat for grounding.
Let your body connect with your mat and release. Breath is natural, let your eyes close and just be.

ASSISTING
Lavender towel and offering a peaceful environment.

SANSKRIT TRANSLATION & STORIES
The name comes from the Sanskrit *shava* "corpse" and *asana* "posture".
There is a colloquial saying that "Shiva without Shakti is Shava" which means that without the power of
action (Shakti), consciousness itself (Shiva) is inactive.

full sequence

INTEGRATION
Child's Pose
Downward Facing Dog
Step Forward, Halfway Lift
Ragdoll
Mountain Pose
Samasthiti
Om x3

AWAKENING
Sun Salutation A - x3
Mountain
Forward Fold
Halfway Lift
High to Low Plank
Upward Facing Dog
Downward Facing Dog
Step Forward, Halfway Lift
Forward Fold

Sun Salutation B - x3
Chair Pose
Forward Fold
Halfway Lift
High to Low Plank
Upward Facing Dog
Downward Facing Dog
Warrior 1, Right side
High Plank to Low Plank
Upward Facing Dog
Downward Facing Dog
Warrior 1, Left side
High to Low Plank
Upward Facing Dog
Downward Facing Dog
End here on third Sun B, or
Step Forward, Halfway Lift
Forward Fold

VITALITY
Crescent Lunge, Right side
Twist to the right
Warrior Two
Extended Side Angle
Side Plank - opposite side
High to Low Plank
Upward Facing Dog
Downward Facing Dog
Crescent Lunge, Left side
Twist to the left
Warrior Two
Extended Side Angle
Side Plank - opposite side
High to Low Plank
Upward Facing Dog
Downward Facing Dog
Step Forward, Halfway Lift
Forward Fold
Chair Pose
Prayer Twist, Right side
Big Toe Pose
Chair Pose
Prayer Twist, Left side
Gorilla
Crow
Vinyasa to Downward Facing Dog
Step Forward, Halfway Lift
Forward fold
Mountain

EQUANIMITY
Eagle - R/L/R/L
Standing Leg Raise, Right side
Extend to the side
Bring back to center
Airplane
Half Moon
Forward Fold
Standing Leg Raise, Left side
Extend to the side
Bring back to center
Airplane
Half Moon
Forward Fold
Dancer - R/L/R/L
Tree - R/L
Mountain
Forward fold
Halfway Lift
High to Low Plank
Upward Facing Dog
Downward Facing Dog

GROUNDING
Warrior Two, Right Side
Triangle
Straddle Fold
Pyramid
Twisting Triangle
Downward Facing Dog
Warrior Two, Left Side
Triangle
Straddle Fold
Pyramid
Twisting Triangle
Downward Facing Dog

IGNITING
High Plank
Lower to the floor
Locust x2
Bow x2
Upward Facing Dog
Downward Facing Dog
Camel
Hero
Camel
Hero
Bridge x2
Wheel x3
Supta Baddha Konasana
Happy Baby

STABILITY
Scissor Legs, Right leg down, pulse x10
Scissor Legs, Left leg down, pulse x10
Legs to the sky
Lower 1/3
Lower another 1/3
Hover 2" from the mat
Legs to the sky
Abdominal Twist x10
Knees to chest
Boat
Downward Facing Dog

OPENING
Half Pigeon, Right side
Double Pigeon
Half Pigeon, Left side
Double Pigeon
Frog

RELEASE
Seated Single Leg Extension - R,L
Seated Forward Bend
Table Top or Incline Plane
Fish

REJUVENATION
Shoulder Stand
Plow Pose
Ear Pressure Pose
* if Head/Handstand, follow with Child's Pose
*Legs Up the Wall optional

DEEP REST
Supine Twist - R/L
Savasana

about the artwork

EARTH ELEMENT
Integration -Malachite

WATER ELEMENT
Awakening - Chrysocolla

FIRE ELEMENTS
Vitality - Ruby
Equanimity - Red Jasper
Grounding - Garnet
Igniting - Rhodochrosite
Stability - Cannelian

AIR ELEMENTS
Opening - Blue Calcite
Release - Blue Topaz
Rejuvenation - Turquoise
Deep Rest - Quartz